Recent Issues o[illegible] uzzy Databases

Dear MyCopy Customer,

This Springer book is a monochrome print version of the eBook to which your library gives you access via SpringerLink. It is available to you at a subsidized price since your library subscribes to at least one Springer eBook subject collection.

Please note that MyCopy books are only offered to library patrons with access to at least one Springer eBook subject collection. MyCopy books are strictly for individual use only.

You may cite this book by referencing the bibliographic data and/or the DOI (Digital Object Identifier) found in the front matter. This book is an exact but monochrome copy of the print version of the eBook on SpringerLink.

Studies in Fuzziness and Soft Computing

Editor-in-chief
Prof. Janusz Kacprzyk
Systems Research Institute
Polish Academy of Sciences
ul. Newelska 6
01-447 Warsaw, Poland
E-mail: kacprzyk@ibspan.waw.pl
http://www.springer.de/cgi-bin/search_book.pl?series=2941

Vol. 3. A. Geyer-Schulz
Fuzzy Rule-Based Expert Systems and Genetic Machine Learning, 2nd ed. 1996
ISBN 3-7908-0964-0

Vol. 4. T. Onisawa and J. Kacprzyk (Eds.)
Reliability and Safety Analyses under Fuzziness, 1995
ISBN 3-7908-0837-7

Vol. 5. P. Bosc and J. Kacprzyk (Eds.)
Fuzziness in Database Management Systems, 1995
ISBN 3-7908-0858-X

Vol. 6. E.S. Lee and Q. Zhu
Fuzzy and Evidence Reasoning, 1995
ISBN 3-7908-0880-6

Vol. 7. B.A. Juliano and W. Bandler
Tracing Chains-of-Thought, 1996
ISBN 3-7908-0922-5

Vol. 8. F. Herrera and J.L. Verdegay (Eds.)
Genetic Algorithms and Soft Computing, 1996
ISBN 3-7908-0956-X

Vol. 9. M. Sato et al.
Fuzzy Clustering Models and Applications, 1997,
ISBN 3-7908-1026-6

Vol. 10. L.C. Jain (Ed.)
Soft Computing Techniques in Knowledge-based Intelligent Engineering Systems, 1997
ISBN 3-7908-1035-5

Vol. 11. W. Mielczarski (Ed.)
Fuzzy Logic Techniques in Power Systems, 1998,
ISBN 3-7908-1044-4

Vol. 12. B. Bouchon-Meunier (Ed.)
Aggregation and Fusion of Imperfect Information, 1998
ISBN 3-7908-1048-7

Vol. 13. E. Orłowska (Ed.)
Incomplete Information: Rough Set Analysis, 1998
ISBN 3-7908-1049-5

Vol. 14. E. Hisdal
Logical Structures for Representation of Knowledge and Uncertainty, 1998
ISBN 3-7908-1056-8

Vol. 15. G.J. Klir and M.J. Wierman
Uncertainty-Based Information, 2nd ed., 1999
ISBN 3-7908-1242-0

Vol. 16. D. Driankov and R. Palm (Eds.)
Advances in Fuzzy Control, 1998
ISBN 3-7908-1090-8

Vol. 17. L. Reznik, V. Dimitrov and J. Kacprzyk (Eds.)
Fuzzy Systems Design, 1998
ISBN 3-7908-1118-1

Vol. 18. L. Polkowski and A. Skowron (Eds.)
Rough Sets in Knowledge Discovery 1, 1998
ISBN 3-7908-1119-X

Vol. 19. L. Polkowski and A. Skowron (Eds.)
Rough Sets in Knowledge Discovery 2, 1998
ISBN 3-7908-1120-3

Vol. 20. J.N. Mordeson and P.S. Nair
Fuzzy Mathematics, 1998
ISBN 3-7908-1121-1

Vol. 21. L.C. Jain and T. Fukuda (Eds.)
Soft Computing for Intelligent Robotic Systems, 1998, ISBN 3-7908-1147-5

Vol. 22. J. Cardoso and H. Camargo (Eds.)
Fuzziness in Petri Nets, 1999
ISBN 3-7908-1158-0

Vol. 23. P.S. Szczepaniak (Ed.)
Computational Intelligence and Applications, 1999
ISBN 3-7908-1161-0

Vol. 24. E. Orłowska (Ed.)
Logic at Work, 1999
ISBN 3-7908-1164-5

continued on page 237

Gloria Bordogna · Gabriella Pasi (Eds.)

Recent Issues on Fuzzy Databases

With 58 Figures
and 15 Tables

Springer-Verlag Berlin Heidelberg GmbH

Dr. Gloria Bordogna
Dr. Gabriella Pasi
Istituto per le Tecnologie Informatiche Multimediali
Consiglio Nazionale delle Ricerche
Via Ampere 56
20133 Milano
Italy
E-mail: gloria.bordogna@itim.mi.cnr.it
gabriella.pasi@itim.mi.cnr.it

DOI 10.1007/978-3-7908-1845-1

Cataloging-in-Publication Data applied for
Die Deutsche Bibliothek – CIP-Einheitsaufnahme
Recent issues on fuzzy databases; with 15 tables / Gloria Bordogna; Gabriella Pasi (ed.). – Heidelberg; New York: Physica-Verl., 2000
(Studies in fuzziness and soft computing; Vol. 53)

Originally published by Physica-Verlag Heidelberg New York in 2000
MyCopy version of the original edition 2000

Hardcover Design: Erich Kirchner, Heidelberg
www.springer.com/mycopy

Foreword

First of all, I would like to congratulate Gabriella Pasi and Gloria Bordogna for the work they accomplished in preparing this new book in the series "*Study in Fuzziness and Soft Computing*". "*Recent Issues on the Management of Fuzziness in Databases*" is undoubtedly a token of their long-lasting and active involvement in the area of Fuzzy Information Retrieval and Fuzzy Database Systems.

This book is really welcome in the area of fuzzy databases where they are not numerous although the first works at the crossroads of fuzzy sets and databases were initiated about twenty years ago by L. Zadeh. Only five books have been published since 1995, when the first volume dedicated to fuzzy databases published in the series "*Study in Fuzziness and Soft Computing*" edited by J. Kacprzyk and myself appeared. Going beyond books strictly speaking, let us also mention the existence of review papers that are part of a couple of handbooks related to fuzzy sets published since 1998.

The area known as fuzzy databases covers a bunch of topics among which:

- flexible queries addressed to regular databases,
- the extension of the notion of a functional dependency,
- data mining and fuzzy summarization,
- querying databases containing imperfect attribute values represented thanks to possibility distributions.

The fuzzy database community has been very active in the past fifteen years and various aspects of fuzzy databases have been dealt with. Significant results have been achieved and reported in conference proceedings or journal papers. Most of these papers are situated along a conceptual/theoretical line and if this type of work is of prime importance and must be maintained, I believe that our community could benefit from two inflexions in the near future. The visibility and recognition of our field of research could be enhanced mainly in two ways, by:

- giving more room to experimental studies, prototypes and implementation issues that have been somewhat neglected so far even if two of the contributions of this volume address such issues; this type of activity is compulsory in order to assess the borders of what is immediately feasible and what is a medium/long range objective,

getting into a closer interaction with the database community; uncertainty has been recognized an important issue in databases for more than ten years (even if probability theory is the priviledged framework) and a recent paper from Ronald Fagin leads me to think that flexibility could be recognized as well very soon; it is then clear that both sides could take advantage of the knowledge acquired separately up to now.

This book gathers ten papers that illustrate some topics being currently a matter of research. They can be categorized into four groups, addressing fairly different aspects. The papers from M. Cobb et al., A. Yazici and K. Akkaya and V. Robinson are concerned with geographic information for which the main issues of imprecision and uncertainty are studied from different perspectives. The papers from G. Chen et al. and J. Kacprzyk and S. Zadrozny put forward two different contributions of fuzzy sets to data mining: the generalization of the notion of an association rule on the one hand and linguistic summaries involving fuzzy terms on the other hand. The topic known as "fuzzy object-oriented database models" is the subject of two papers by de Tré et al. and Bordogna et al. The first one describes a model which is in the spirit of the recommendations of the ODMG, whereas the other one focuses on implementation issues. The last three papers are devoted to different extensions of relational database systems. I. Blanco et al. discuss the implementation of a fuzzy deductive database system, M. Nakata investigates the extension of the division of relations when the universal quantifier is changed into a weaker one (such as "most") and P. Bosc et al. propose an overview of the various types of queries that are possible in presence of ill-known attribute values.

"Recent research issues on the management of fuzziness in databases" provides a view on different aspects of the area of fuzzy databases which is potentially useful for a wide range of readers, from Master or Ph. D students to researchers and engineers. Let us thank again the instigators of this valuable contribution.

Patrick BOSC
Lannion, France
January 21, 2000

Preface

With the diffusion of wide-area networks, the information available on-line has increased, and consequently the need for the development of effective and flexible information systems has raised a great emphasis on the research in this field.
By flexibility is meant the ability of the systems to manage different kinds of information and to adapt their behaviour to the context of the application and the users.
An important aspect of flexibility is the ability to manage the information characterized by some kind of imperfection, which indeed affects the data involved in many real applications. The word imperfection has been introduced to indicate one of the following faults of information: imprecision, vagueness, uncertainty, and inconsistency. Imperfection in a database may characterize either the data themselves, or the user queries or both.
In the literature, a unifying approach to manage imperfect information through fuzzy set theory and possibility theory has been proposed and widely adopted to generalize the data paradigms.
This book focuses on the recent research issues on the application of fuzzy set theory to extend the functionalities of data base management systems.
During the past five years, the research in this field has moved from a merely theoretical framework mainly focused on the definition of fuzzy extensions of the relational database model to the consideration of other database paradigms also in relation with their application in specific contexts. The semantic and the deductive database models have been extended within fuzzy sets and possibility theory and implementations of these models have been proposed and used in specific contexts such as Geographic Information System applications.
While the first fuzzy theoretical extensions of relational databases have considered the definition of flexible query languages addressed to precise data and successively the representation and management of imperfect data, in the recent years other topics have been tackled such as for example fuzzy data mining and summarization and fuzzy functional dependencies definitions.
In the present book we have collected some contributions focusing on these up-to-date topics.
In particular we have grouped the papers into four parts:

Part I. Fuzzy Querying of Databases: it groups two papers dealing with new aspects pertaining the definition of flexible query languages. The paper by *Patrick Bosc, Laurence Duval, and Olivier Pivert* titled *"Value-Based and Representation-Based Querying of Possibilistic Databases"* after making a clear state of the art on flexible querying to databases identifies a novel

semantics for querying imperfect data: the query conditions are regarded as constraints on the data representations.
The second paper, by *Michinori Nakata* titled *"Flexible Division Operators in Possibility-based Fuzzy Relational Databases"* extends the classical division operation based on the notion of fuzzy linguistic quantifiers.

Part II. Data Mining based on Fuzzy Sets collects two contributions: the first one by *Guoqing Chen, Qiang Wei, and Etienne E. Kerre* titled *"Fuzzy Data Mining: Discovery of Fuzzy Generalized Association Rules"* proposes a fuzzy extension of the classical method based on association rules to mine generalized association rules in presence of fuzzy taxonomies.
The second contribution by *Janusz Kacprzyk and Slawomir Zadrozny* titled *"On Combining Intelligent Querying and Data Mining Using Fuzzy Logic Concepts"* proposes an approach to the production of linguistic summaries of a database based on the extension of the notion of association rules through fuzzy terms and linguistic quantifiers; they make use of the FQUERY language defined for the *access* package to support the generation of database linguistic summaries.

Part III. Geographic Information Systems Dealing with Uncertainty gathers three contributions: the first one, by *Maria Cobb, Harold Foley, Frederick Petry, and Kevin Shaw* titled *"Uncertainty in Distributed and Interoperable Spatial Information Systems"* tackles the problem of the management of spatial data from different sources in presence of uncertainty and proposes solutions to data incompatibilities based on fuzzy set.
The second contribution by *Vincent B. Robinson* titled *"On Fuzzy Sets and the Management of Uncertainty in an Intelligent Geographic Information System"* addresses the issues of the representation and management of uncertainty at both the landscape analysis and query levels by fuzzy sets within the Knowledge Based Land Information Manager and Simulator.
The third contribution by *Adnan Yazici and Kemal Akkaya* titled *"Conceptual Modeling of Geographic Information System Applications"* uses ExIFO_2, a semantic data model incorporating a graph-based representation framework, to model the uncertainty related to geographic objects and their relationships, and then maps the conceptual model into the Fuzzy Object Oriented Database.

Part IV. Fuzzy Database Models: Definition and Implementation comprehends three contributions: the first one by *Guy de Tré, Rita de Caluwe, and Bart Van der Cruyssen* titled *"A Generalised Object-Oriented Database Model"* proposes a formal framework for defining an ODMG-compliant object oriented database model dealing with fuzzy and uncertain information. The notion of generalised constraints is used for enforcing integrity rules and for defining the semantics of the database.
The second contribution by *Ignacio Blanco, Juan C. Cubero, Olga Pons, and*

Amparo Vila titled *"An Implementation for Fuzzy Deductive Relational Databases"* proposes a model to integrate the representation and management of deductive rules into a fuzzy relational database model. New data can be deduced from both precise and imprecise data.
The last contribution by *Gloria Bordogna, Alberto Leporati, Dario Lucarella, and Gabriella Pasi* titled *"The Fuzzy Object Oriented Database Management System"*, after defining the Fuzzy Object Oriented Database model which makes use of a graph-based paradigm for representing both the database schema and the instances, describes the characteristics of a prototypal system implementation.

To conclude, the goal of this book is to give an up-to-date overview of the current research on fuzzy databases and to contribute to stimulate the interest in this field.
We think that the research carried out so far has matured some fuzzy extensions of classical databases that are feasible to be implemented and applied fruitfully in real applications.

Acknowledgements

First of all we would like to thank the authors of the papers. It was a pleasure reviewing and editing their interesting contributions.
Our most sincere gratitude also goes to Patrick Bosc for his stimulating foreword.
We would also like to express our gratitude to the anonimous referees that reviewed the papers and substantially contributed to the quality of the book.
Finally we would like to thank the series editor of *"Studies in Fuzziness and Soft Computing"* Janusz Kacprzyk, for proposing the preparation of this volume.

Gloria BORDOGNA and Gabriella PASI
Milano, Italy
February 14, 2000

Contents

Part IV. Fuzzy Database Models: Definition and Implementation

Part I.

Fuzzy Querying of Databases

Value-Based and Representation-Based Querying of Possibilistic Databases

Patrick Bosc [1], Laurence Duval[2], and Olivier Pivert [1]

[1] IRISA/ENSSAT, Technopole Anticipa BP 447, 22305 Lannion Cedex, France
[2] IRISA/ENSAI, Campus de Ker Lann, 35170 Bruz, France

Abstract. In this paper, we address the issue of querying imperfect data represented by possibility distributions. We distinguish between two types of queries: those involving conditions on the values, and those involving criteria on the representations of ill-known values. These two approaches are successively considered. We first recall some classical results relating to the querying of databases involving null values, and we point out the problems that arise in the specific context of disjunctive weighted data when value-based queries are dealt with. The necessity of defining a typology of relevant queries is emphasized. Then, we introduce a new querying framework allowing to handle ill-known data at a representation level. This framework, based on the notion of a weighted set, offers an alternative solution to the use of value-based queries and thus could be used to extend the querying capabilities of database systems aimed at handling ill-known values.

Keywords. Relational databases, ill-known values, uncertainty, query language

1. Introduction

For many years, database management systems have been used almost exclusively for applications (mainly business applications) where data were perfectly known. However, in real life, information is often imperfect in several ways. Giving a precise characterization of imperfectness is not easy, but several different situations can be pointed out, notably the two following ones.

First, attribute values can be ill-known. Indeed, the attribute value of an item may be completely unknown, partially known (i.e. known as belonging to a proper subset of the attribute domain), or somewhat uncertain. This latter case can be dealt with through different formalisms (probability [1, 2], and possibility theory [3,4], in particular; see [5] for a survey). Besides, an attribute may be irrelevant for some of the considered items; moreover we may not know whether the value does not exist or is simply unknown.

Second, the information relating to a fact can be uncertain. In this case, weighted relations may be handled, and the information in a tuple is qualified with a certainty (or a possibility, or a probability, depending on the type of knowledge available) equal to its weight.

In this paper, we only consider the first situation, i.e., databases involving ill-known attribute values and we study the special case where ill-known data are represented by possibility distributions. Besides the representation issue, an important aspect is the handling of such data. Up to now, the querying of databases containing imperfect values has been considered to relate to the same general principle as usual database querying: the conditions allowed concern the *value* that a variable (associated with a given attribute) can take. In the first part of the paper, we consider this value-based querying approach and we point out different problems which arise in this context. As it is well known since the works on null values (cf. for instance [6]), when data are ill-known, value-based queries lead to uncertainty in the answers since the event "item t satisfies condition c" becomes itself uncertain. For instance, let us consider a database describing houses and suppose that the price of house H1 is represented by the disjunctive set {$80,000, $100,000, $120,000}. An example of a usual filtering (i.e., value-based) criterion is: "find the houses whose price is higher than $90,000" (in this case, H1 belongs to the set of possible answers, but not to the set of certain answers because of the candidate $80,000). More generally, a relation containing ill-known values can be seen as a set of possible worlds. A value-based query has then to be evaluated against each possible world and the result is itself a set of possible worlds. We show that in the context of databases containing uncertain data represented by disjunctive weighted sets (e.g., possibility distributions), three classes of value-based queries rise out. A first class corresponds to selection-projection queries: the conditions concern individual tuples and do not imply to explicitely compute sets of possible worlds. A second class gathers queries that involve a comparison of ill-known values from several relations. The result of such a query is an explicit set of possible worlds. Queries of the third class check up on the possibility (resp. certainty) for a given tuple to belong to the result of a query of the first or second class. In the first part of the paper, we point out the problems raised by these different classes of queries in the context of disjunctive weighted data.

The value-based approach corresponds to an important type of information need, but some filtering criteria of another type can be envisaged, which prove themselves useful too in the context of a database containing ill-known values. For instance, one could think of querying the database mentioned above by means of filtering criteria such as: "find the houses whose price representations contain at least two possible values (i.e., the houses whose prices are ill-known)". In this case, H1 belongs to the result, which is no longer uncertain. Such an example illustrates a querying approach where a condition concerns the *set of candidates* for a given attribute value, i.e., the *representation* of this attribute value. The approach we propose goes beyond the handling of simple disjunctions: it is also intended to deal with weighted disjunctions in order to allow the querying of possibilistic data. Then, the information concerning the underlying levels of imprecision and/or uncertainty has to be taken into account too. An important point is that representation-based conditions cannot be expressed using the same constructs as value-based conditions. Therefore, there is a need for a specific framework suitable for this new type of query, and a notable contribution of this paper resides in the introduction of such a framework.

The remainder of this paper is structured as follows. Section 2 is devoted to the value-based querying of databases containing ill-known attribute values. We point out the fact that imperfect data can be represented in terms of different kinds of disjunctions and we study the value-based queries that make sense in this context. In Section 3, we introduce the basic principles of the representation-based querying approach. Section 4 describes a framework for the expression of representation-based conditions and illustrates the use of this framework through several examples of queries.

2. Value-based querying of databases containing imperfect data

The necessity of dealing with imperfect information comprises two aspects: the description of the data stored, and the handling of such data, i.e., the querying aspect. Hereafter, we first present several approaches that have been proposed to model incomplete data. Let us recall that we only consider imperfection in atomic data values. Approaches like maybe tuples and disjunctive databases are not described here since they concern imperfection in associations between data (see for instance [5] for a survey). As far as the querying aspect is concerned, an important question concerns the types of queries that make sense in the context of imperfect data. In particular, one may wonder whether usual queries are still relevant, both from a tractability point of view and a user point of view (the question, in the latter case, being the possibility of easily interpreting the results). We refer to these usual queries as "value-based queries" insofar as they involve conditions aimed at comparing a variable (i.e., an attribute value) with a constant, or two attribute values with one another. In order to point out the problems that arise in a context of imperfect data, we first recall some classical results obtained in the framework of databases with nulls, then we consider the case where imperfect data are represented as possibility distributions.

2.1 Representing imperfect values: a brief survey

2.1.1 Null Values

For databases under the closed world assumption, two notable types of null values have been proposed [7, 8, 9]: i) existential value (denoted ω): the value exists, but it is not known, ii) inexistent value (denoted $\perp$): the value does not exist (in other words, it is not applicable). The most well-known type of relations containing null values are the Codd relations. Imielinsky and Lipski [6] introduced several unknown values for obtaining a richer modeling power, and they generalized the relational algebra operators. Marked null values are represented by variables, where two different variables may or may not represent the same constant.

2.1.2 Disjunctive sets

Or-sets (or disjunctive sets) [10, 11] generalize existential null values. An existential value represents an attribute whose actual value is in a database domain. An or-set represents an attribute whose value is in an explicit, smaller set.

An ordinary atomic value can be viewed as a singleton. An existential value corresponds to an or-set containing the entire attribute domain (it is indeed a disjunction of all the values of the domain).

2.1.3 Weighted disjunctive sets

Probability theory and possibility theory are two possible frameworks for managing uncertainty. In both frameworks, an uncertain value is represented by a weighted disjunction, the difference residing in the meaning of the weights and in the nature of the uncertainty modeled.

A probabilistic database (see for instance [1, 12, 13]) contains probabilistic information about data values in the following way. Probabilistic relations have key attributes that are deterministic, as in classical relations. The other attributes may be deterministic or stochastic. The latter are described with the help of probabilistic sets. An example of such a probabilistic relation is shown hereafter:

Takes	Student	Course
	John	{Algebra/0.5, Calculus/0.4}
	Ann	{Physics/0.5, Calculus/0.5}

Let us notice that, in the first tuple, the probabilistic weights do not sum up to 1, which means that John can also take no course. Then, the relation above has six possible worlds, and a probability degree can be computed for each world.

On the other hand, the possibility theory [14] provides a purely ordinal model for uncertainty where imprecision is represented by means of a preference relation coded by a total order over the possible situations. This constitutes the fundamental difference between this theory and the probability theory which is quantitative. The concept of a possibility distribution in a universe X refers to an application π: X $\rightarrow$ [0, 1] which is interpreted as a restriction of the possible values of a variable taking its values in X. We denote $\pi(a)$ the degree of possibility that the effective value of x is a. The possibilitic approach can be applied for representing uncertain values in a relational database in the following way (see [3, 4, 15]). The available information about the value of a single-valued attribute A for a tuple x is represented by a possibility distribution $\pi_{A(x)}$ on $D \cup \{e\}$ where D denotes the domain of attribute A and e is an extra-element which stands for the case when the attribute does not apply to x. If information is consistent, there should exist a value in $D \cup \{e\}$ for A(x), which leads to the normalization condition $\max_d \pi_{A(x)}(d) = 1$ (i.e., at least one value in $D \cup \{e\}$ is completely possible). This approach proposes a unified framework for representing precise values, as well as imprecise ones (regular sets) or vague ones (fuzzy sets), and various null value situations.

2.2 Value-based querying of databases containing nulls: some classical results

The issue of querying incomplete databases was notably studied by Lipski [10, 11] in the context of databases containing existential null values. The landmark paper [6] laid the formal groundwork for incomplete databases with nulls of the unknown kind and introduced the notion of a representation system. Since then, there has been considerable work on querying incomplete information databases. Surveys about incomplete databases and queries can be found in [16, 17] and [5]. The following presentation is mainly based on [17].

2.2.1 Basic notions

At the semantic level, an incomplete database T is a set of possible (complete) databases, one of which corresponds to the true state of the world. This set of possible databases is usually denoted by rep(T). What is now the answer to a query addressed to an incomplete database? In the general case, we get a set of possible answers, each computed from one possible state of the real world (this set is called the elementwise extension of the query in [6]). Thus the answer is another incomplete database, just as the answer to a relational query in the complete case is a relation. This set of all possible answers is called the exact answer. The set of tuples that belong to every world of the exact answer is called the certain answer (or sure answer). The set of tuples that belong to at least one world of the exact answer is called the maybe answer. Note that in the maybe answer some of the tuples come from different possible answers.

As noted above, the simplest way of modeling an incomplete relation is a Codd table (or table for short), i.e., a relation with constants and variables, in which no variable occurs twice. In this model, a possible world corresponds to a certain valuation of the variables. It is important to note that, in general, the exact answer cannot be represented as a table. The problem lies in the weakness of the representation system of tables. The following example, drawn from [17], illustrates this problem.

Example 1. Let us consider the following Codd relation:

R	A	B	C
	0	1	x
	y	z	1
	2	0	v

whose examples of corresponding instances (i.e., possible states of this relation) are:

R	A	B	C
	0	1	2
	2	0	1
	2	0	0

I_1

R	A	B	C
	0	1	2
	3	0	1
	2	0	5

I_2

R	A	B	C
	0	1	2
	2	0	1
	2	0	0

I_3

R	A	B	C
	0	1	1
	2	0	1

I_4

Consider the algebraic query $\sigma_{A=3}$ (T). There is no table representing the exact answer to this query. A possible answer (e.g., for I_1) is the empty relation, whereas there are nonempty possible answers (e.g., for I_2). Suppose that there exists a table

T' representing the set of possible answers. Either T' is empty and $\sigma_{A=3}$ (T) is not in rep(T'); or T' is nonempty and the empty relation is not in rep(T'). This is a contradiction, so no such T' can exist♦

A first approach is then to be less demanding and to use weak representation systems [17]. A second approach is to consider richer models leading to a complete representation system for all of the relational algebra operations. A third approach consists in considering queries of a new kind, i.e., queries with "good properties".

2.2.2 Weak representation systems

In this approach, one no longer requires that the answer to a query be a representation of the exact answer. Instead, one asks which are the tuples that are surely in the answer (sure answer). Similarly, one may ask for the tuples that are possibly in the answer (maybe answer). In [6], it is shown that tables form a weak representation system for selection-projection (SP) queries, i.e., relational algebra limited to selection (involving only equalities and inequalities) and projection. If union or join are added, it has been shown that tables no longer form a weak representation system.

An improvement consists in using so-called naive tables, i.e., tables where variables may repeat. It has been shown in [6] that naive tables (also called v-tables) form a weak representation system for relational algebra restricted to positive conditions.

2.2.3 Conditional tables

As we have seen, Codd tables and naive tables do not allow to represent the exact answer to a relational query in general, due to the fact that the presence or absence of certain tuples in a possible answer is conditioned by certain properties of the valuation. To capture this, it has been proposed in [6] to extend the representation with conditions on variables, which yields conditional tables (also called c-tables). A condition is a conjunct of equality atoms of the form $x = y$, $x = c$ and of inequality atoms of the form $x \neq y$, $x \neq c$, where x and y are variables and c is a constant. Conditions may be associated with table T in two ways: (1) a global condition Φ_T is associated with the entire table; (2) a local condition φ_t is associated with one tuple t of table T. An example of a c-table is:

T	A	B	
	$x \neq 2, y \neq 2$		
	0	1	
	1	x	$y \neq 0$
	y	x	$x = y$

and some possible instances (obtained by valuating x, y to (0, 0), (0, 1) and (1, 0), respectively) are:

J_1	A	B
	0	1
	0	0

J_2	A	B
	0	1
	1	0

J_3	A	B
	0	1

It has been shown in [6] that conditional tables form a strong representation system for the relational algebra.

2.2.4 Complexity aspects

The complexity aspect is twofold: it concerns both the interpretation of the result from the user's side and the tractability of the evaluation. Obviously, the answer to a usual query in the form of a conditional table is not easily interpretable. It is not a realistic solution either to provide the user with all the possible instances of the result. Moreover, this would be untractable because of the potentially huge number of worlds to compute. Let us consider for instance a join query involving two relations, each containing one ill-known attribute. Let us assume that three candidates are possible for each ill-known attribute value, that each relation contains 1000 tuples, and that the join condition applies to the ill-known attribute in each relation. The number of possible worlds to be computed is 3^{2000}.

An alternative solution consists in considering queries of a different kind, based on the notion of membership (or inclusion). A typical example is: is a set of tuples possibly (or certainly) in the answer to query q? The complexity of such queries has been studied by Abiteboul et al. in [18]. For a query q that can be evaluated in polynomial time on complete databases, deciding whether relation I is a possible answer, or whether I is a set of possible tuples (i.e., a maybe answer) can be answered in NP; checking whether I is a set of certain tuples is in CO-NP.

2.3 Value-based querying of possibilistic databases

In this section, we consider the case where ill-known values are represented in the form of an explicit (possibly weighted) distribution of values.

2.3.1 Selection-projection queries

Let us first consider the case where ill-known data are represented by or-sets. In this context, an ill-known value is represented by a nonweighted set containing all the possible candidates for the given attribute value. Let us emphasize that we consider only "usual" or-sets, i.e., *explicit* disjunctive sets of values, and let us recall that this framework does not allow to represent that two or-sets are the same (we do not consider here more sophisticated models such as or-relations, described in [19]).

Example 2. Let us consider a military database containing aerial images of aircrafts (each image is supposed to represent a single aircraft), described by the set of attributes: (#id, location, date, type). The attributes #id, location, and date are supposed to take precise values whereas the attribute "type" describing the type of aircraft present in the picture will generally take imperfect values (in many cases, during the scene interpretation process, uncertainty will result from ambiguities in interpretations). An example of content is given hereafter:

#id	location	date	type
7	Dolon	02-03-96	{Mig-31, Mig-29, Mig-23}
9	Krasnoyark	01-12-92	{Su-27, Su-30, Mig-29}
17	Arkhangelsk	09-27-96	Mig-31
22	Dolon	03-02-98	{Su-27, Su-30}

♦

Or-sets form a weak representation system for selection-projection queries. From a complexity point a view, an interesting result is that selection and projection do not necessitate to compute explicitely all the different possible worlds, but that such queries can be evaluated on the "compact" model itself in order to get the maybe answer and the sure answer. In this context, the set of maybe tuples resulting from a selection-projection query is richer than the strict maybe answer inasmuch as the initial representation of each possibly satisfactory tuple is kept, including the candidates that do not fulfill the selection condition (whereas the strict maybe answer is the union of the satisfactory *instances* of the initial tuples). It is also worth noticing that the different possible answers that constitute the exact answer can be computed from the two relations obtained: a possible answer is the union of the sure answer and any combination of the tuples of the maybe answer (including the empty set case).

Example 3. Let us consider the extension above and the query: "find the pictures representing a Mig-31". The maybe answer is computed the following way: for each tuple, one checks whether Mig-31 belongs to the set of candidate values for attribute "type" and, if it is the case, the tuple is returned. Here, the maybe answer contains the tuples corresponding to images 7 and 17. On the other hand, a tuple belongs to the sure answer if the only possible value specified for attribute "type" is Mig-31. Here, the sure answer contains image #17 only♦

Keeping the initial representations of ill-known data in the resulting relation has the following consequence: conjunctive selection conditions involving nonindependent predicates cannot be decomposed. In other words, if predicates P1 and P2 are nonindependent, we do not have in general: select(R, P1 and P2) = select(select(R, P1), P2).

Example 4. Let us consider the following relation R:

R	#id	A	B
	1	8	{6, 9}

Let us consider the conjunctive selection condition: S = (A $\leq$ B and B $\neq$ 9). The result of the query: select(R, S) should be the empty relation since no candidate value for attribute B makes the condition true. On the other hand, the result of the query: select(select(R, A $\leq$ B), B $\neq$ 9) is equal to relation R ♦

In the possibilistic framework, the situation is rather similar. The main difference is that the result of a selection query is a twofold weighted relation where each tuple is provided with two degrees expressing the extent to which the given tuple is a possible (resp. certain) answer. The projection consists in removing the duplicates and aggregating the degrees by means of the max operation. For the sake of clarity, we will see the result of a selection-projection query as a pair of fuzzy relations. Let us denote by ΠP (resp. NP) the fuzzy relation that contains the items that possibly (resp. necessarily) satisfies condition P.

Considering a relation $R(A_1,...,A_n)$, we take $A_i(x)$ to be the possibility distribution representing the value of attribute A_i in the tuple x and the predicate P representing a criterion applicable to A_i. The membership degree of a tuple x to ΠP and NP are respectively given by [4]:

$$\mu_{\Pi P}(x) = \Pi(P; A_i(x)) = \sup_{u \in dom(A_i)} \min(\pi_{A_i}(u), \mu_P(u)) \quad (1)$$

$$\mu_{NP}(x) = N(P; A_i(x)) = 1 - \Pi(\overline{P}; A(x)) = \inf_{u \in dom(A_i)} \max(1 - \pi_{A_i}(u), \mu_P(u)) \quad (2).$$

$\Pi(P; A_i(x))$ estimates the extent to which there is a value restricted by $\pi_A(x)$ compatible with P and $N(P; A_i(x))$ the extent to which all the values more or less possible for $A_i(x)$ are included in P (the condition P being seen here as a crisp set of acceptable values). Function μ_P is such that $\mu_P(d) = 1$ if P(d) is true, 0 otherwise.

Example 5. Let us illustrate the matching mechanism described above. Consider the query: "find the images of recent model aircrafts" involving the predicate P = "recent-model" defined as:

$P(x) \equiv [x \in \{$Mig-33, Su-37, Ka-52, Su-33, Mig-31, IL-76, Mig-29$\}]$.

Let us consider the image described as:

I = (3, Dolon, 04-23-96, {1/Mig-31, .8/Mig-29, .6/Mig-23, .6/Su-27, .4/Su-33}).

The evaluation of the predicate "image I represents a recent-model aircraft" will be based on the computation of the values:

$\Pi(F \mid I.type) = \sup(\min(1, 1), \min(.8, 1), \min(.6, 0), \min(.6, 0), \min(.4, 1)) = 1$

and

$N(F \mid I.type) = \inf(\max(0, 1), \max(.2, 1), \max(.4, 0), \max(.4, 0), \max(.6, 1)) = .4$♦

Compound conditions involving disjunction, conjunction or negation of elementary conditions can be handled using some basic relations of possibility theory, which express decomposability properties of possibility and necessity degrees with respect to conjunctions and disjunctions provided that the attribute values are logically independent. During the projection, duplicate removal consists in computing the maximum of the degrees.

Again, one can see that in the possibilistic framework, selection-projection queries do not impose to compute explicitely all the different possible worlds, but can be evaluated on the "compact" model itself. Moreover, these queries are perfectly tractable (their data complexity is linear). Of course, the Projection-Selection-Join (PSJ) queries such that the join condition applies to precisely known attributes have the same good properties.

2.3.2 Joins and queries involving set-oriented operations

Queries involving two relations, such as join queries (in the general case where the join attributes may take ill-known values) or queries including set-oriented operators, are more problematical. Basically, the problem stems from the nonindependence of tuples in the result. As a matter of fact, pairs of tuples cannot be considered independently during the evaluation since a same tuple of the first relation has to be compared to several tuples of the second relation. Thus, the need for a coherent result implies to take explicitely into account "valuations" at a database level and not at a tuple level (in this context, a "valuation" corresponds to the choice of one candidate for each ill-known value).

First, let us point out that relations containing explicit disjunctions (e.g. possibilistic relations) do not form a strong representation system for Cartesian product queries (and consequently for join queries).

Example 6. Let us consider the two following relations:

R1

X	Y
{a, b}	c

R2

X	Z
{a, d}	e
b	f

The exact answer to the query (Cartesian product): R1 × R2 is made of four possible worlds:

X	Y	X	Z
a	c	a	e
a	c	b	f

X	Y	X	Z
a	c	d	e
a	c	b	f

X	Y	X	Z
b	c	a	e
b	c	b	f

X	Y	X	Z
b	c	d	e
b	c	b	f

Then, the exact answer to the join query R1[X = X]R2 gathers three distinct possible answers: the empty relation Ø and the two following nonempty relations:

X	Y	Z
a	c	e

X	Y	Z
b	c	f

This exact answer cannot be represented by a single relation such as the initial ones♦

Let us also mention that such relations do not even form a weak representation system for Projection-Selection-Join-Union (PSJU) since the notions of maybe and certain answers are not compatible with compositionality of PSJU queries.

Besides the representation issue, one has to face the same difficulties as those encountered in the case of nulls, i.e., the high complexity of the evaluation and the impossibility for the user to interpret a result expressed as a (potentially huge) set of possible worlds. Obviously, these problems remain the same when one deals with possibility distributions.

As far as intersection and difference operations are concerned, the basic problem is the same (we do not go into the detail here for the sake of brevity). The important point is that such queries are in general intractable and that their results are not manageable by the user.

2.3.3 Possibility and necessity queries

Considering the problems mentioned above, it does not seem realistic to allow for relational queries other than selection-projection in the context of a database containing disjunctive attribute values. Nevertheless, one can think of some basic computational questions about such incomplete information databases, in the same spirit as those suggested in [18] in the framework of databases with nulls. Consider a database B containing disjunctive attribute values, a query q (which can involve any algebraic operators), a (regular) relation I, and a tuple t. Some typical yes-no questions are the following:

(q-Membership) Is I a possible answer for q and B?
(q-Possibility) Is t possibly in the answer?
(q-Certainty) Is t certainly in the answer?

In the possibilistic framework, the exact answer to a PSJU query is a set of regular relations, each provided with a possibility degree equal to the minimum of the degrees attached to the corresponding tuple valuations. The corresponding questions are:

(q-Membership) Is I a somewhat possible answer for q and B?

and the result is then the possibility degree attached to the corresponding possible relation in the exact answer to q if such a relation exists (it is 0 otherwise);

(q-Possibility) To what extent is t a possible answer?

In this case, the result is the maximum of the possibility degrees attached to the possible relations of the exact answer containing t;

(q-Certainty) To what extent is t a certain answer?

The result corresponds then to the necessity that t belongs to all the possible relations of the exact answer. It is equal to 1 minus the possibility that there exists a possible result relation not containing t, i.e. it is 1 minus the maximum of the possibility degrees attached to the possible result relations that do not contain t.

Example 7. Let us consider the following possibilistic relation:

R	#id	location	date	type
	7	Dolon	02-03-96	{1/Mig-31, .8Mig-29, .3Mig-23}
	9	Krasnoyark	01-12-92	{1/Su-27, .6/Mig-29, .4/Mig-23}
	17	Arkhangelsk	09-27-96	1/Mig-31
	22	Dolon	03-02-98	{1/Su-27, .5/Su-30}

and the possibility question: to what extent is it possible that image 7 and image 9 represent the same airplane? In other words: does the pair (7, 9) belong to the result of query q herafter:

q = project(join(R, R, type = type), #id1, #id2)?

It is interesting to notice that the evaluation process can be optimized. Let us consider now query q' hereafter:

q' = join(project(select(R, #id = 7), type), project(select(R, #id = 9), type),
type1 = type2).

The possibility query then becomes: to what extent is it possible for the answer to q' to be nonempty? There are three candidates for the attribute "type" in both images, thus the exact answer contains nine relations. Among these nine, seven are empty (with the respective degrees 1, .6, .4, .8, .4, .3, .3). The two remaining ones contain respectively Mig-29 (R1) and Mig-23 (R2). The possibility degrees associated with these relations are .6 for R1 and .3 for R2. Thus, the answer to the preceding possibility question is .6. The corresponding necessity question would yield a degree zero since it is completely possible for the answer to be empty ♦

As mentioned before, Abiteboul et al. have studied the complexity of such queries in the framework of databases involving null values [18]. A similar work should be done in the possibilistic framework in order to determine whether these queries are interesting in practice, i.e., if they can be evaluated in a reasonable time. Intuitively, one may think that the results reported in [18] constitute upper bounds of the data complexities that can be expected in the possibilistic framework insofar as possibility distributions are more specific (i.e., "less incomplete") than null values. More precisely, it seems interesting to envisage some optimization mechanisms based on rewriting (in the same spirit as described in the preceding example), the basic idea being to use the available attribue values of the target tuple(s) (the values 7 and 9 in the example) as constraints that can be introduced into the underlying query, in order to reduce the number of possible worlds to be computed.

2.4 Conclusion

As it is well known, the crucial difference between complete and incomplete information is the large number of possible valuations for the latter case. The number of such valuations may grow exponentially in the input size. As a consequence, some relational queries which are perfectly trivial when complete data are handled become highly problematical in the context of incomplete data. Here, we have pointed out the specific forms that this problem takes in the case where incomplete information is represented by means of possibility distributions. It appears that two solutions are possible: either one limits oneself to "simple" queries, i.e., selection-projection queries, or one considers queries of a different type, e.g., possibility/certainty queries (whose data complexity may be high, but these queries have the good quality of leading to yes-no answers and thus do not entail any difficulty in interpreting the result). In the following, we study an alternative type of queries whose evaluation does not raise any particular problem of complexity: queries based on the representations of ill-known values.

3. A new type of queries: representation-based queries

3.1 Principles of representation-based queries

Representation-based querying [20] consists in exploiting at a query level some concepts of the data model which concern the representation of imprecision/uncertainty. In order to illustrate this approach, let us consider again a database containing incomplete descriptions of aerial images, such as that described in Section 2.

First let us assume that imperfect values are represented by or-sets. For instance, let us suppose that image I1 represents an airplane whose (ill-known) type is represented by the disjunctive set {Mig-31, Mig-23, Su-33}. An example of a usual (i.e., value-based) query is: "find the images representing a Mig" (in this case, I1 belongs to the set of possible answers, but not to the set of certain answers because of the candidate Su-33). On the other hand, an example of a representation-based query is: "find the images for which at least two different sorts of Mig are possible for the airplane type". In this case, I1 belongs to the result, which is no longer uncertain. In this querying approach, a condition concerns the *representation* of a given attribute value, i.e., the *set of candidates* for that attribute value. In this sense, there is a certain similarity between representation-based querying and the querying of databases containing multivalued attributes. In both cases, a condition may refer explicitely to a set of values, the difference being that in the first case, the elements are seen as exclusive candidate values whereas in the latter case, they are all supposed to be actual values.

It is important to notice that representation-based queries are not just value-based queries expressed another way, but that they are queries of a different nature. A value-based criterion applying to an ill-known value has to be evaluated on each possible world associated with the attribute value (even though the explicit computation of those worlds is not always necessary). Now let us consider the query: "find the images for which the set of type candidates contains both a Mig and a Sukhoi". The value-based approach is not suitable for the expression of such a query which, intrinsically, does not rest on a concept of possibility of matching and thus does not induce any uncertainty. In this particular example, it would be possible, though, to use a value-based formulation, but that would be quite unnatural (one would have to require the set of images that possibly represent a Mig, then the set of images that possibly represent a Sukhoi, and then to compute the intersection of these two sets). What is even more, some representation-based queries do not make sense from a value-based point of view (for instance the query: "find the images representing a Mig and not a Sukhoi" is almost meaningless in the value-based approach since an airplane cannot be a Mig and a Sukhoi at the same time).

The approach we propose is intended to deal with weighted disjunctions in order to allow the querying of possibilistic data. Then, the information concerning the underlying levels of imprecision and/or uncertainty has to be taken into account. Let us consider again the aircraft example and the possibilistic tuple: (7, Dolon, 02-03-96, {1/Mig-31, .8/Mig-29, .6/Mig-23, .4/Su-33}). An example of a representation-based query in this context is: "find the images where all the Mig candidates are possible over .5" (then, the image above belongs to the result).

A remark concerns the types of operation that are meaningful in a representation-based querying framework. A representation-based query is basically a matter of selection (and projection) insofar as most relevant representation-based conditions concern a single relation (which can be the result of another represented-based query or the result of a value-based query involving conditions on usual attribute values only). However we will see in the following that other types of representation-based operations (i.e., join, intersection, union, difference) can be useful as well.

3.2 Motivations

The main motivation underlying the representation-based querying approach is to be able to exploit *at a query level* all the information available concerning the qualification of imperfectness in the data. In other words, we want to be able to express conditions on the *descriptions* of ill-known data. The concept of a representation-based query is somewhat similar to that of a structural query introduced in [21] in a slightly different context (the authors consider complex objects represented as unweighted sets (disjunctive or conjunctive) whose elements can be themselves regular or disjunctive sets). Representation-based queries can notably be used to:

i) express conditions on specified sets of candidates (the specified set being a subset of a distribution representing an ill-known attribute value). The generic query is: "find the tuples such that all the elements of a specified subset of the candidate values (for a certain attribute) satisfy a given condition".

ii) compute aggregates on the weighted sets corresponding to the representations of ill-known data (e.g., the cardinality of a specified subset of candidate values for a given attribute) and to use these aggregates inside conditions.

iii) compare a piece of data with a given vague pattern. Let us consider for instance the case of a database system containing facial images such as described in [22]. Such images can be described by various facial features such as length of hair, size of eyes, length of nose, etc. It is not quite easy to determine a precise value for each of these attributes, and a more convenient way of describing the images (and of formulating queries) consists in using linguistic labels (interpreted as possibility distributions) such as "short", "narrow", "fairly long", etc. In the representation-based querying framework, the comparison is based on the notion of synonymy of representations, contrary to the value-based framework where the comparison is founded on the notion of possibility/necessity of matching.

3.3 Some examples of relevant representation-based conditions

3.3.1 Conditions involving one representation

Let A be a weighted set corresponding to an attribute value, and V the domain of A. Let v (resp. d) denote a variable on V (resp. on D = [0, 1]). Representation-based conditions involving one representation can refer to:

i) the degrees of some values of the domain. For instance:

- the value a_1 is preferred to the value a_2 for A
- the values a_1 and ... and a_n are possible over a threshold α for A
- a_1 is (one of) the most possible value(s) for A

ii) the cardinality of a given α-cut. For instance:

- at most n values are possible over α for A

iii) the imprecision of an attribute value. For instance:

- the attribute value A is not precisely known.

3.3.2 Conditions involving two representations

i) selection (the query involves an "ideal" representation M, expressed either as an explicit weighted set or as a linguistic label). For instance:

- A is equal to M (strict equality or approximate equality, cf. Section 4.2)
- A is interchangeable with M (assuming a resemblance relation on the domain is available)
- A is more specific (resp. more general) than M.

ii) representation-based join, union, intersection or difference. The following predicate illustrates the concept of a representation-based self-join condition:

- there exist two tuples with the "same" representation for attribute A.

3.4 Specificities of these queries and the need for an appropriate framework

As noted before, Prade and Testemale [4] have defined some extended (value-based) relational operators allowing to manipulate relations including ill-known values. The selection operation is based on a matching mechanism which delivers a twofold fuzzy relation, i.e., a relation where each tuple t is provided with two degrees Π and N expressing the possibility (resp. necessity) of its membership to the relation (cf. Section 2). The purpose of this section is to show that the expression power of the value-based querying framework is too limited to allow the expression of all kinds of representation-based queries. Some representation-based conditions can be expressed in the value-based querying framework though, but at the expense of a greater complexity in the formulation.

Let us consider a relation R(X, A), X and A being sets of attributes, X being the key of relation R. Using an SQL-like language where the result of a selection is a twofold fuzzy relation including two additional attributes Π and N, we will try to express some queries considered in section 3.3 in the value-based querying framework.

i) the value a_1 is preferred to the value a_2 for attribute A

select * **from** R **into** R1 **where** A = a_1;
select * **from** R **into** R2 **where** A = a_2;
select R1.X, R1.A **from** R1, R2 **where** R1.X = R2.X **and** R1.$\Pi \geq$ R2.Π;

In relation R1 (resp. R2), the additional attribute Π indicates the extent to which the A-value of a tuple is possibly equal to a_1 (resp. a_2). Then, the last query retrieves the tuples whose A-value is more possibly equal to a_1 than to a_2.

ii) all of the values $\{a_1, ..., a_n\}$ are possible over a degree α for attribute A

select * **from** R **into** R1 **where** A = a_1;
...
select * **from** R **into** Rn **where** A = a_n;
select R1.X, R1.A **from** R1, ..., Rn
where R1.X = R2.X **and** ... **and** R1.X = Rn.X
and R1.$\Pi \geq \alpha$ **and** ... **and** Rn.$\Pi \geq \alpha$;

In relation R1 (resp. R2, ..., Rn), the additional attribute Π indicates the extent to which the A-value of a tuple is possibly equal to a_1 (resp a_2, ..., a_n). Then, the last query retrieves the tuples that possibly belong to R1 and ... and to Rn at a degree greater or equal to α.

iii) a_1 is the only value completely possible for attribute A

> **select** * **from** R **into** R1 **where** A = a_1;
> **select** * **from** R **into** R2 **where** A ≠ a_1;
> **select** R1.X, R1.A **from** R1, R2
> **where** R1.X = R2.X **and** R1.Π = 1 **and** R2.Π ≠ 1;

In relation R1 (resp. R2), the additional attribute Π indicates the extent to which the A-value of a tuple is possibly equal to a_1 (resp. is possibly different from a_1). Then, the last query retrieves the tuples such that a_1 is a completely possible value for attribute A whereas it is not the case for any other value.

iv) at most n values are possible over a degree α for attribute A

> not expressible.

v) the value of attribute A is ill-known

> not expressible.

Let us consider now the second class of representation-based queries, i.e., queries involving a "syntactic" comparison of two possibility distributions. Their evaluation can be based on any measure of similarity of fuzzy sets. Obviously, these queries are not expressible in the regular possibilistic framework, where the comparison of ill-known values is based on fuzzy pattern matching, since their evaluation is not a matter of possibility and necessity, but a matter of "distance" between two representations.

This brief comparison shows that the expression of representation-based conditions in the regular possibilistic framework is problematical: either these conditions require multiple queries, or they are not expressible at all. This result justifies our proposal of a specific querying framework suitable for this kind of selection conditions. This framework, based on the notion of a weighted set, is described hereafter.

4. A language for the expression of representation-based conditions

4.1 Conditions involving one representation

Two kinds of constructs are necessary in order to express conditions on sets of values. On the one hand, one must have some functions extracting a set from another one. On the other hand, aggregate functions must be available. The standard aggregate functions are: count, min, max, avg and sum. It must be noticed that the last four ones make sense only if the considered set is defined on an ordered domain, and the last two only if the domain is numerical.

Now let us consider how these constructs can be applied to weighted sets (e.g., possibility distributions). A weighted set is assumed to be a set of pairs (v_i, d_i) defined on $V \times D$ where V denotes the domain of the considered attribute and D is the unit interval. Let us notice that or-sets are a special case of a possibility distribution where all the elements have the degree 1.

In this context, an extraction function f is defined on $V \times D$ and it consists of a conjunction of two conditions: one used to filter the values, the other one to filter the degrees:

$$f = [f_v, f_d] \text{ with } f_v: V \rightarrow \{\text{true, false}\} \text{ and } f_d: D \rightarrow \{\text{true, false}\}.$$

A condition $f_v(x)$ is of the form:

$$f_v(x) = x\ \theta_1\ v_1\ \textbf{and/or} \ldots \textbf{and/or}\ x\ \theta_n\ v_n$$

where the θ_i's are comparators ($\{=, \neq, >, \geq, <, \leq\}$ for numerical attributes, $\{=, \neq,$ contains, etc$\}$ for character strings and so on), and the v_i's are elements of V (either literals or aggregate expressions, this latter case corresponding to a nested query). Similarly, a condition $f_d(y)$ is of the form:

$$f_d(y) = y\ \theta_1\ d_1\ \textbf{and/or} \ldots \textbf{and/or}\ y\ \theta_n\ d_n$$

where the θ_i's are comparators ($=, \neq, >, \geq, <, \leq$), and the d_i's are elements of D (either literals or aggregate expressions, this latter case corresponding again to a nested query).

As far as aggregates are concerned, one needs to consider functions aggregating values of V and functions aggregating degrees. In other terms, one needs to have available two sets of standard aggregate functions:

$$G_v = \{\text{v-count, v-id, v-min, v-max, v-avg, v-sum}\}$$
$$G_d = \{\text{d-count, d-id, d-min, d-max, d-avg, d-sum}\}.$$

The aggregate function v-id (resp. d-id) is assumed to return the value of a singleton given as an argument (and d-id returns 0 if its argument is the empty set). As noted before, depending on V, only a subset of G_v may make sense. Similarly, depending on the framework used — probabilistic or possibilistic — only a subset of G_d makes sense. In addition to these standard functions, some user-defined aggregate functions may also appear in a query (cf. example iii in the next section).

Finally, the general form of a representation-based condition C: $W \rightarrow \{\text{true, false}\}$ (where W is the set of all the weighted sets defined on $V \times D$) will be:

$$C(A) = g_1(f_1(A))\ \theta_1\ \alpha_1 \text{ and/or} \ldots \text{and/or } g_n(f_n(A))\ \theta_n\ \alpha_n$$

where $g_i \in G_v \cup G_d$, the f_i's are extraction functions as defined above, the θ_i's are comparators and the α_i's are either literals or aggregate expressions.

The querying framework based on these principles appears quite simple, but, as we will see in the next section, it reveals itself powerful enough to allow to formulate many kinds of representation-based queries.

4.2 Some examples in the context of a database describing images

Let us consider again the query classification outlined in Section 3.3. Hereafter, we extend this classification and give some examples of representation-based queries involving one representation expressed in the framework proposed above. Let A be a possibility distribution, and V the corresponding domain. Let v (resp. d) denote a variable on V (resp. on D = [0, 1]). Again, we consider as an example a military database containing aerial images of aircrafts.

i) Conditions referring to the degrees of some values of the domain:

- find the images which represent more likely a Mig-29 than a Mig-23

  C(A) = (d-id([v = 'Mig-29', true](A)) ≥ d-id([v = 'Mig-23', true](A)))

- find the images such that all the candidates which are possible over .3 are of the type Mig

  C(A) = (v-count([true, d > .3](A)) = v-count([v = 'Mig-*', d > .3](A))).

ii) Conditions referring to the cardinality of a given α-cut (i.e., set of values whose grades are not less than α):

- find the images for which at most 2 types of airplane are considered possible over .3

  C(A) = (v-count([true, d ≥ .3](A)) ≤ 2)

- find the images for which the only best candidate is 'Mig-29'

  C(A) = (d-id([v = 'Mig-29', true](A) = 1 and
  v-count([true, d = 1](A) = 1).

iii) Conditions on the imprecision of an attribute value.

- find the images representing airplanes whose type is not precisely known (i.e., there are more than one candidate)

  C(A) = (v-count([true, true](A)) > 1)

If one wants to qualify more accurately the amount of imprecision attached to an attribute value, one can use a specific measure indicating the extent to which a value is imprecise. Such measures are studied in [23]. Let us denote by d-spec such a function. It can play the role of a user-defined aggregate in a query, which allows to express a filtering predicate such as:

- the amount of imprecision associated to A is less than α:

$$C(A) = (\text{d-spec}([\text{true}, \text{true}](A)) < \alpha.$$

iv) Conditions on the uncertainty level attached to a given attribute value: it is the largest value α in [0, 1] such that $\forall x \in V$, the degree associated to x in the considered distribution is not less than α. In order to express such queries, it is necessary to have available the cardinality of the related domain (denoted |domain(A)| hereafter).

- the uncertainty level associated to A is less than λ

$$C(A) = (\text{v-count}([\text{true}, d \geq \lambda](A) < |\text{domain}(A)|)$$

Example 8. Let us consider an item I from the part of the database where images of helicopters are stored: I = (17, Arkhangelsk, 09-27-96, {1/Ka-52, .7/Ka-50, .4/Mi-26T}), and let us assume that the underlying domain is D = {Ka-50, Ka-52, Mi-26T}. The uncertainty level associated to the attribute "type" of image I is equal to .4 ♦

4.3 Conditions involving two representations

Several methods have been proposed to compare possibility distributions or fuzzy sets and one can distinguish among two families of approaches. In the first family, a measure is used to evaluate the possibility degree of (approximate) equality between two imprecise values [24, 25, 4]. In the second family, what is measured is the extent to which two *representations* are globally close to each other [26, 27, 28, 29]. In the representation-based querying framework, it is quite clear that only the second family of approach makes sense.

Let us consider an attribute A and two items x and y whose A-values are ill-known. Let us denote $\pi_{A(x)}$ and $\pi_{A(y)}$ the possibility distributions that we want to compare. Let D be the domain of attribute A. First, let us recall the expression of the strict equality:

$$\forall d \in D, \pi_{A(x)}(d) = \pi_{A(y)}(d).$$

Several authors have proposed to relax the preceding measure into a measure of approximate equality. In [29], Raju and Majumdar define the fuzzy equality measure, denoted EQ, in the following way:

$$\mu_{EQ}(\pi_{A(x)}, \pi_{A(y)}) = \min_{u \in D} \psi(\pi_{A(x)}(u), \pi_{A(y)}(u))$$

where ψ is a resemblance relation (i.e., reflexive and symmetric) over [0, 1]. Similar methods have also been advocated in [28] and [4]. An alternate approach consists in defining the similarity of two fuzzy sets (two possibility distributions in our case) A and B as a function of $A \cap B$, $B - A$ and $A - B$. This approach is studied in particular in [27] where different kinds of measures of comparison are considered. In [22] and [30], a representation-based approach founded on a so-called fuzziness dependent matching measure is proposed in order to allow for a flexible retrieval of facial images from a database containing linguistic descriptions.

These approaches, which provide a spectrum of methods to compare ill-know values, are suitable when strict equality is the only way to compare domain values. However, there are many circumstances which reveal the necessity of specifying a resemblance relation on domains. In this case, the comparison consists in estimating the extent to which the respective representations of A(x) and A(y) are interchangeable with respect to the considered resemblance relation. In this spirit, a measure of interchangeability, generalizing Raju and Majumdar's measure of fuzzy equality, is defined in [26]. Its principle is briefly described hereafter. Let us denote by Res the resemblance relation expressing fuzzy equality between the values of domain D. The interchangeability degree related to the pair (A(x), A(y)) with respect to Res is the degree to which A(x) can be replaced with A(y) and reciprocally:

$$\mu_{INT}(A(x), A(y)) = \min(\mu_{repl}(A(x), A(y)), \mu_{repl}(A(y), A(x))).$$

An imprecise value A(x) can be replaced with another imprecise value A(y) if, for each representative $<u/\pi_{A(x)}(u)>$ of A(x), there exists a representative $<v/\pi_{A(y)}(v)>$ of A(y) close to $<u/\pi_{A(x)}(u)>$. The degree of closeness between such two representatives is expressed as an aggregation of the degree of closeness between u and v on the one hand, and the degree of closeness between $\pi_{A(x)}(u)$ and $\pi_{A(y)}(v)$ on the other hand. The influence of a representative $<u/\pi_{A(x)}(u)>$ on the global degree $\mu_{repl}(A(x), (A(y))$ is all the more important as $\pi_{A(x)}(u)$ is high, which yields:

$$\mu_{repl}(A(x), A(y)) = \inf_{u \in support(A(x))} \max(1 - \pi_{A(x)}(u), \mu_{s(A(x), A(y))}(u))$$

with:

$$\mu_{s(A(x), A(y))}(u) = \sup_{v \in support(A(y))} \min(\mu_{Res}(u, v), \mu_\theta(\pi_{A(x)}(u), \pi_{A(y)}(v))).$$

In this formula, θ is a proximity measure over [0, 1] which plays the same role as ψ in Raju-Majumdar's approach.

It must be noticed that as soon as a measure of approximate equality (or interchangeability) is used, the result is a fuzzy relation, i.e., a relation where each tuple is provided with a grade in the unit interval expressing its membership to the relation.

Let us now consider the practical use of such representation-based measures. First, they can be used, obviously, to compare two ill-known attribute values. Let us consider two ill-known data D1 and D2. In the usual fuzzy querying framework, the comparison of D1 and D2 relies on the fuzzy pattern matching method (cf. Section 2), which consists in computing the possibility (resp. necessity) degree of the condition "D1 = D2". In the representation-based querying framework, D1 and D2 are viewed as linguistic labels and what we measure is the approximate synonymy of their respective representations.

Example 9. Let us consider the following description of an image I1:

I1 = (27, Krasnoyarsk, 01-12-92, {1/Su-27, 1/Su-30, .7/Mig-29, .2/Yak-130})

Let us consider the query: "find the pictures taken over Krasnoyarsk in 1992 representing an airplane similar to the one in image I1" and let us assume that the database contains notably the following description:

I2 = (51, Krasnoyarsk, 04-15-92, {1/Su-30, .9/Mig-29, .8/Su-27, .4/Mig-23}).

If the strict equality was used, it is clear that image I1 would not belong to the result. Using Raju-Majumdar's measure of approximate equality with $\psi(a, b) = 1 - |a - b|$, the matching degree between I1 and I2 is equal to:

μ_{EQ}(I1.type, I2.type)
= min(.8 (Su-27), 1 (Su-30), .8 (Mig-29), .8 (Yak-130), .6 (Mig-23))
= .6.

Using the measure of interchangeability described above, along with the resemblance relation Res given hereafter:

	Su-30	Su-27	Mig-29	Mig-23	Yak-130
Su-30	1	.8	.3	.1	0
Su-27		1	.7	.2	.1
Mig-29			1	.7	.3
Mig-23				1	.8
Yak-130					1

Table 1. Resemblance relation on attribute "type"

and $\theta(a, b) = 1 - |a - b|$, the evaluation requires the following calculations:

μ_{repl}(I1.type, I2.type) = inf(max(0, x), max(0, y), max(.3, z), max(.8, t)) with:

x = $\mu_{s(I1.type,\ I2.type)}$(Su-27)
= sup(min(.8, 1), min(.7, .9), min(1, .8), min(.2, .4) = .8

y = $\mu_{s(I1.type,\ I2.type)}$(Su-30)
= sup(min(1, 1), min(.3, .9), min(.8, .8), min(.1, .4)) = 1

z = $\mu_{s(I1.type,\ I2.type)}$(Mig-29)
= sup(min(.3, .7), min(1, .8), min(.7, .9), min(.7, .7)) = .8

t = $\mu_{s(I1.type,\ I2.type)}$(Yak-130)
= sup(min(0, .2), min(.3, .3), min(.1, .4), min(.8, .8)) = .8

thus: μ_{repl}(I1.type, I2.type) = .8

μ_{repl}(I2.type, I1.type) = inf(max(0, x'), max(.1, y'), max(.2, z'), max(.6, t')) with:

x' = $\mu_{s(I2.type,\ I1.type)}$(Su-30)
= sup(min(.8, 1), min(1, 1), min(.3, .7), min(0, .2)) = 1
y' = $\mu_{s(I2.type,\ I1.type)}$(Mig-29)
= sup(min(.7, .9), min(.3, .9), min(1, .8), min(.3, .3)) = .8
z' = $\mu_{s(I2.type,\ I1.type)}$(Su-27)
= sup(min(1, .8), min(.8, .8), min(.7, .9), min(.1, .4)) = .8
t' = $\mu_{s(I2.type,\ I1.type)}$(Mig-23)
= sup(min(.2, .4), min(.1, .4), min(.7, .7), min(.8, .8)) = .8

thus: μ_{repl}(I2.type, I1.type) = .8

and finally: μ_{INT}(I1.type, I2.type) = min(.8, .8) = .8 ♦

On the other hand, these measures can be used to compare an ill-known attribute value D with a fuzzy predicate P. The basic idea is the same: one evaluates the extent to which the predicate and the value represent the same concept. For example, let us consider a possibility distribution D representing John's age and a fuzzy predicate P = "middle-aged" (represented by a fuzzy set). While the fuzzy pattern matching method allows to measure the extent to which John is possibly (resp. necessary) middle-aged, the querying approach based on representations can be used to measure the extent to which the description of John's age and the fuzzy set representing the concept "middle-aged" are close to each other. The value computed will then correspond to the synonymy of the representation of John's age and "middle-aged". This approach is especially useful in the context of applications where user queries can be conveniently expressed by means of linguistic terms defined on continuous domains. This is the case for instance in the application described in [22] which deals with the retrieval of facial images, and where a user can submit a query such as "find the images of persons with fairly long hair, a slender face and a high forehead".

Lastly, the concept of representation-based comparison can be used to define the notions of representation-based intersection, union and difference in a straightforward manner.

5. Conclusion

In this paper, we have addressed the issue of querying databases that involve imperfect values represented by possibility distributions. We have shown that in this context, one can distinguish between two types of queries: those involving conditions on the values, and those involving criteria that apply to the representations of the values. These two approaches have been successively considered. First, we have pointed out the different problems inherent to the value-based querying approach in the particular context of disjunctive weighted data. The necessity of defining a typology of relevant queries has been emphasized. Some future works on this issue should concern the study of efficient evaluation mechanisms, especially for the so-called possibility/necessity queries that have only been studied in a context of null values so far. In the second part of the paper, we have proposed a querying framework allowing to handle possibilistic data at a representation level. This framework, based on the notion of a weighted set, involves three types of constructs (extraction function, aggregate, comparison operator), and reveals itself simple and powerful. It is worth mentioning that the general principle of representation-based querying is not strongly linked to the possibilistic framework, but could be applied to probabilistic data too. Future works should consist in: i) generalizing this framework in order to deal with gradual conditions, ii) defining a higher level query language integrating the functionalities described here.

References

1. Barbará, D., Garcia-Molina, H., Porter, D.: The management of probabilistic data. IEEE Transactions on Knowledge and Data Engineering **4** (1992) 487-502
2. Dey, D., Sarkar, S.: A probabilistic relational model and algebra. ACM Transactions on Database Systems **21** (1996) 339-369
3. Prade, H.: Lipski's approach to incomplete information data bases restated and generalized in the setting of Zadeh's possibility theory. Information Systems **9** (1984) 27-42
4. Prade, H., Testemale, C.: Generalizing database relational algebra for the treatment of incomplete/uncertain information and vague queries. Information Sciences **34** (1984) 115-143
5. Zimanyi, E., Pirotte, A.: Imperfect information in relational databases. In: Motro, A., Smets, P. (eds.): Uncertainty Management in Information Systems: From Needs to Solutions, Kluwer Academic Publishers (1997) 35-87
6. Imielinski, T., Lipski, W.: Incomplete information in relational databases. Journal of the ACM **31** (1984) 761-791
7. Codd, E.F.: Extending the Database Relational Model to Capture More Meaning. ACM Transactions on Database Systems **4** (1979) 397-434
8. Codd, E.F.: Missing Information (applicable and inapplicable) in relational databases. SIGMOD Record **15** (1986) 53-78
9. Biskup, J.: A formal approach to null values in database relations. In: Gallaire, H., Nicolas, J.M. (eds.): Advances in Data Base Theory, Vol. 1. Plenum Press (1981) 299-341
10. Lipski, W.: On semantic issues connected with incomplete information databases. ACM Transactions on Database Systems **4** (1979) 262-296
11. Lipski, W.: On databases with incomplete information. Journal of the ACM **28** (1981) 41-70
12. Lee, S.K.: An Extended Relational Database Model for Uncertain and Imprecise Information. Proc. of the 18th VLDB Conference, Vancouver, Canada (1992) 211-220

13. Lakshaman, L.S.V., Leone, N., Ross, R., Subrahmanian, V.S.: Probview: A flexible probabilistic database system. ACM Transactions on Database Systems **22** (1997) 419-469
14. Zadeh, L.A.: Fuzzy sets as a basis for a theory of possibility. Fuzzy Sets and Systems **1** (1978) 3-28
15. Bosc, P., Prade, H.: An Introduction to the Fuzzy Set and Possibility Theory-Based Treatment of Flexible Queries and Uncertain or Imprecise Databases. In: Motro, A., Smets, P. (eds.): Uncertainty Management in Information Systems. Kluwer Academic Publishers (1997) 285-324
16. Grahne, G.: The problem of incomplete information in relational databases. Lecture Notes in Computer Science, Vol. 554. Springer Verlag, Berlin (1991)
17. Abiteboul, S., Hull, R., Vianu, V.: Foundations of databases. Addison-Wesley (1995)
18. Abiteboul, S., Kanellakis, P., Grahne, G.: On the representation and querying of sets of possible worlds. Theoretical Computer Science **78** (1991) 159-187
19. Imielinski, T., Vadaparty, K.: Complexity of query processing in databases with or-objects. Proc. of the 8th ACM SIGACT-SIGMOD Symposium on Principles of Database Systems (1989) 51-65
20. Bosc, P., Pivert, O.: A new approach to the filtering of ill-known data. Proc. of the 7th IEEE International Conference on Fuzzy Systems, Anchorage, USA (1998) 1302-1307
21. Libkin, L., Wong, L.: Semantic representations and query languages for or-sets. Proc. of the 12th PODS Conference (1993) 37-48
22. Gasos, J., Ralescu, A.: Adapting query representation to improve retrieval in a fuzzy database. International Journal of Uncertainty, Fuzziness and Knowledge-Based Systems **3** (1995) 57-77
23. Yager, R.R.: On the specificity of a possibility distribution. In: Dubois, D., Prade, H., Yager, R.R. (eds.): Readings in Fuzzy Sets for Intelligent Systems, Morgan Kaufmann Publishers (1993) 203-216
24. Chen, G.Q., Kerre, E.E., Vandenbulcke, J.: A general treatment of data redundancy in a fuzzy relational data model. Journal of the American Society for Information Science **43** (1992) 304-311
25. Cubero, J.C., Vila, M.A.: A new definition of fuzzy functional dependency in fuzzy relational databases. Journal of Intelligent Systems **9** (1994) 441-448
26. Bosc, P., Pivert, O.: On the comparison of imprecise values in fuzzy databases. Proc of the 6th IEEE International Conference on Fuzzy Systems (1997) 707-712
27. Bouchon-Meunier, B., Rifqi, M., Bothorel, S.: Towards general measures of comparison of objects. Fuzzy Sets and Systems **84** (1996) 143-153
28. Liu, W.: The fuzzy functional dependency on the basis of the semantic distance. Fuzzy Sets and Systems **59** (1993) 173-179
29. Raju, K.V.S.V.N., Majumdar, A.K.: Fuzzy functional dependencies and lossless join decomposition of fuzzy relational database systems. ACM Transactions on Database Systems **13** (1988) 129-166
30. Fukushima, S., Ralescu, A.: Improved retrieval in a fuzzy database from adjusted user input. Journal of Intelligent Information Systems **5** (1995) 183-210

Flexible Division Operators in Possibility-based Fuzzy Relational Databases*

Michinori Nakata

Department of Information Science, Chiba-Keizai College
4-3-30 Todoroki-cho, Inage-ku, Chiba 263-0021 JAPAN
nakata@chiba-kc.ac.jp

Abstract. A division operator, which corresponds to the universal quantifier *for all*, is extended to apply it to a variety of situations; namely, to cases where quantifiers such as *a few*, *several*, *at most about seven*, *most* and so on are used. The extension that has done so far is limited to a case that quantifiers, like *exists*, *at least k*, and *most*, are expressed by a non-decreasing membership function. An approach proposed here has not the restriction. Moreover, the extension is made under a possibility-based fuzzy relational model based on necessity and possibility measures. Therefore, the approach is more general and more realistic than the ones that have been proposed so far.

Keywords: Division operators, Relational algebra, Quantifiers, Fuzzy databases, Relational databases

1 Introduction

A division operator is one of operators that compose relational algebra in relational models[1]. The operator is useful for a kind of queries; for example, *Select suppliers that supply all parts required for product A.* The operator is the expression that corresponds to the universal quantifier *for all.*

Several investigators have extended this operator to use it in fuzzy relational models[2–12]. This is made from two viewpoints. One is to extend the division operator so that we can use it in fuzzy relational databases[3, 4, 6, 8, 9, 11, 12]. In this case, the extension is made by considering the following three points: a membership attribute value that each tuple has[4–6, 8, 11], a resemblance degree between domain elements[3, 12], and imprecise attribute values expressed by a possibility distribution[7, 12]. The other extension is to relax *for all* so that a division operator can be flexibly applied to a variety of situations[2, 5, 8, 11]. The division operator is called a flexible one. This

*This paper is a revised and slightly expanded version of a paper entitled "Flexible division operators in fuzzy databases" by the same author, prepared for Vietnam-Japan Bilateral Symposium on Fuzzy Systems and Applications (VJFUZZY '98), Halong Bay, Vietnam, September 30-October 2, 1998 and appearing on pp. 334-341 in Proceedings.

paper is related with the flexible division operator.

This kind of extensions have been so far made under a restriction that quantifiers used in place of *for all* have a non-decreasing membership function, which are called a non-decreasing type of quantifiers; for example, *at least k*, *most*, and so on. However, there are many familiar quantifiers that cannot be expressed by a non-decreasing membership function such as *a few*, *several*, and *at most seven*. This paper gives a formulation to make a division operator flexible without any restrictions for a membership function by which a quantifier is expressed.

Flexible division operators have been so far developed only in the framework of a simple fuzzy relational model where every attribute value is precise and each membership attribute value in a relation is a single value in the interval $[0, 1]$. However, extensions of relational models have been usually proposed under allowing imprecise information and using fuzzy restrictions by fuzzy sets and possibility theory (for example, see references [13–17]). Thus, the formulation of flexible division operators is addressed in a fuzzy relational model allowing imprecise attribute values expressed by a possibility distribution and using fuzzy restrictions, which is called a possibility-based fuzzy relational model.

This paper is organized as follows. In Section 2, a possibility-based fuzzy relational model is expressed, which is based on necessity and possibility measures. In Section 3, first, a formulation of a division operator is briefly surveyed to use it in the possibility-based fuzzy relational model, and then we make the division operator flexible. Section 4 is concluding remarks.

2 Possibility-based fuzzy relational models

Fuzzy relational models[13–17] are an extended version of conventional relational models in the framework of fuzzy sets and possibility theory that are proposed by Zadeh[18–20]. Fuzzy relational scheme $R(\mathcal{A}_R, \mu)$, different from relational scheme in that a membership attribute appears, consists of a set of ordinary attributes $\mathcal{A}_R = \{A_1, A_2, \ldots, A_n\}$ and a membership attribute μ, as is expressed by some authors[21–23]. All data are expressed by a possibility distribution. Formally, value $t[A_l]$ of attribute A_l in tuple t is represented by a normal possibility distribution.

$$t[A_l] = \sum_{u \in dom(A_l)} \pi_{t[A_l]}(u)/u,$$

$$\sup_{u \in dom(A_l)} \pi_{t[A_l]}(u) = 1,$$

where $\sum_{u \in dom(A_l)}$ stands for the union over $u \in dom(A_l)$ and each domain $dom(A_l)$ is a crisp set.

When an attribute value of a tuple in a relation is represented by a possibility distribution, possible values that the attribute value can take are elements composing that possibility distribution. A possible value corresponds to a possible tuple value on the attribute. Hence, an element in an attribute value can be related to a possible tuple value on that attribute, as is shown in the following example.

Example
In the first tuple of relation *pp* possible values that attribute $PRICE$ can take are elements 20 and 40 that compose a possibility distribution $0.8/20+1/40$.[†] These elements correspond to possible tuple values $\langle z, 20\rangle$ and $\langle z, 40\rangle$, respectively.

pp

$PART$	$PRICE$	μ
$1/z$	$0.8/20+1/40$	1,1
$1/x$	$1/30$	1,1

As a result of allowing imprecise attribute values expressed by a possibility distribution, the most fundamental component is a possible value, as is shown in the example.

A relation r is an extended set of tuple values with their membership attribute value $t_r[\mu] = (t_r[\mu_N], t_r[\mu_\Pi])$, where $t_r[\mu_N]$ (resp. $t_r[\mu_\Pi]$) is a degree in necessity(resp. in possibility) to which tuple value $t_r[\mathcal{A}_R]$ belongs to r.

$$r = \{\langle t_r[\mathcal{A}_R], (t_r[\mu_N], t_r[\mu_\Pi])\rangle \mid t_r[\mu_\Pi] > 0\}$$

Some authors adopt fuzzy relations with given weights as base relations[21–23]. These weights are interpreted as uncertainty of bonds among attribute values, degree of importance, degree of uncertainty, degree of fulfillment, truth value, and so on[8, 17, 22]. Which interpretation is the best depends on circumstances. Other authors use relations without weights as base relations, although derived relations have two weights based on necessity and possibility measures as a result of query processing[15]. A base relation and derived relations from it must have the same properties; otherwise, we cannot apply the same operations to the base relation and the derived relations.

Relations are made under restrictions that consist of constraints in integrity constraints and/or specified conditions in query processing. A degree to which a tuple value belongs to a relation is a compatibility degree of that tuple value with imposed restrictions on that relation in our fuzzy relational model. The degree is not given, but calculated by using necessity and possibility measures. For the sake of simplicity, suppose that restriction c imposed on relation r is "A_i is a." When a is expressed by membership function μ_a, compatibility degree $\Pi(c \mid t_r[\mathcal{A}_R])$ (resp. $N(c \mid t_r[\mathcal{A}_R])$) of tuple value $t_r[\mathcal{A}_R]$ in possibility (resp. in necessity) with c is calculated according to Prade and

[†] + denotes the union rather than the arithmetic sum.

Testemale[15]:

$$\begin{aligned}\Pi(c \mid t_r[\mathcal{A}_R]) &= \Pi(t_r[A_i] = a) \\ &= \sup_{u \in D_i} \min(\pi_{t_r[A_i]}(u), \mu_a(u)), \\ N(c \mid t_r[\mathcal{A}_R]) &= 1 - \Pi(\neg(t_r[A_i] = a)) \\ &= \inf_{u \in D_i} \max(1 - \pi_{t_r[A_i]}(u), \mu_a(u)).\end{aligned}$$

If there are no restrictions imposed on relation r except this restriction c, a membership attribute value of tuple t_r is:

$$\begin{aligned}t_r[\mu] &= (t_r[\mu_N], t_r[\mu_\Pi]) \\ &= (N(c \mid t_r[\mathcal{A}_R]), \Pi(c \mid t_r[\mathcal{A}_R])).\end{aligned}$$

Example

We suppose that a fuzzy restriction "*PRICE is from about* 35 *to* 90" is imposed on relation *pp* in the previous example. Compatibility degrees of values of attribute *PRICE* with the restriction are calculated. Let *from about* 35 *to* 90 be expressed by a membership function $\{0.3/20, 0.7/25, 1/30, 1/35, 1/40, 1/45, \cdots, 1/90\}$. For the first tuple compatibility degrees are 0.3 in necessity and 1 in possibility, respectively. For the second tuple these degrees are equal to 1. We obtain the following relation under the restriction.

pp

PART	*PRICE*	μ
$1/z$	$0.8/20 + 1/40$	0.3,1
$1/x$	$1/30$	1,1

As is shown in the above example, a membership attribute is necessarily required under allowing imprecise attribute values and/or using fuzzy restrictions. This must be contained in the framework of fuzzy relational models, which is justification for a fuzzy relational scheme to contain the membership attribute.

We use both possibility and necessity measures, although some authors use only possibility measures. We cannot always distinguish data by using only possibility measures under accepting fuzzy restrictions, as is shown in the previous example where two tuples have the same compatibility degree in possibility with the restriction. Thus, we use necessity measures along with possibility measures. The previous example shows that two tuples can be distinguished by using necessity measures. This justifies using a pair of values as a value of membership attribute. The membership attribute values in all relations are calculated by necessity and possibility measures. Thus, all the information obtained in our fuzzy relational model is based on necessity and possibility measures; base and derived relations have the same properties.

3 Division operators

A division operator usable in the possibility-based fuzzy relational model mentioned in section 2 has to be formulated. This formulation is already obtained from the three viewpoints addressed in section 1. Hence, a limited description of the formulation is given for the next subsection (see reference[12] for details). And then we weaken the universal quantifier *for all.* This leads to making division operators flexible.

3.1 Extension of division operators

Recently a division operator has been defined by several researchers in fuzzy relational databases[2–12]. Suppose that r and s are relations defined on relational schemes $R(\mathcal{A}_R)$ and $S(\mathcal{A}_S)$, respectively. According to Ullman[24] in relational models the operator is defined by derived relation $r \div s$ that includes subtuple t of r if for every tuple $t_s \in s$ tuple $\langle t, t_s \rangle$ exists in r; namely,

$$r \div s = \{t \mid \forall t_s \in s,\ \langle t, t_s \rangle \in r\},$$

where $\langle t, t_s \rangle[\mathcal{A}_{R-S}] = t$ and $\langle t, t_s \rangle[\mathcal{A}_S] = t_s[\mathcal{A}_S]$ for $\langle t, t_s \rangle$.

Dubois and Prade have extended this definition in a simplified fuzzy relational model based on fuzzy relations with single-valued membership attribute values and precise values of attributes[8]. According to them the above definition is extended in the possibility-based fuzzy relational model addressed in section 2[12]. When r and s are relations defined on fuzzy relational schemes $R(\mathcal{A}_R, \mu)$ and $S(\mathcal{A}_S, \mu)$, respectively, membership attribute value $t_{r\div s}[\mu]$ of tuple value $t_{r\div s}[\mathcal{A}_{R-S}]$ equal to t in derived relation $r \div s$ is defined:

$$t_{r\div s}[\mu] = (t_{r\div s}[\mu_N], t_{r\div s}[\mu_\Pi]), \tag{1}$$

$$t_{r\div s}[\mu_\Pi] = \min_{t_s}(t_s[\mu] \rightarrow \langle t, t_s \rangle_r[\mu])_\Pi, \tag{2}$$

$$t_{r\div s}[\mu_N] = \min_{t_s}(t_s[\mu] \rightarrow \langle t, t_s \rangle_r[\mu])_N. \tag{3}$$

where $\rightarrow$ is a fuzzy implication, $\langle t, t_s \rangle_r$ is a tuple with $\langle t, t_s \rangle_r[\mathcal{A}_{R-S}] = t$ and $\langle t, t_s \rangle_r[\mathcal{A}_S] = t_s[\mathcal{A}_S]$ in relation r, and $\langle t, t_s \rangle_r[\mu]$ is a membership attribute value attached to tuple value $\langle t, t_s \rangle_r[\mathcal{A}_R]$.

How the fuzzy implication is evaluated depends on requirements specified to division operators. Detailed discussion for the topic have been already had in division operators (see references [9, 11, 12] and so on). In the possibility-based fuzzy relational model described in section 2, the membership attribute values are calculated by necessity and possibility measures. They are compatibility degrees in necessity and in possibility of a tuple value with restrictions. Thus, suitable fuzzy implications would be ones that do not contain the four rules of arithmetic for these degrees, because, for example, it seems to be difficult to interpret a difference between two degrees in possibility or in necessity. Representative usable implications are Dienes implication and Gödel implication that are typical examples of S-implication and R-implication, respectively and do not contain the four rules of arithmetic. Here, we show the formulation using Dienes implication (see reference [12] for Gödel implication), and then

$$(t_s[\mu] \to \langle t, t_s\rangle_r[\mu])_N = \max(1 - t_s[\mu_\Pi], \langle t, t_s\rangle_r[\mu_N]), \quad (4)$$

$$(t_s[\mu] \to \langle t, t_s\rangle_r[\mu])_\Pi = \max(1 - t_s[\mu_N], \langle t, t_s\rangle_r[\mu_\Pi]). \quad (5)$$

When imprecise attribute values expressed by a possibility distribution are allowed[12],[‡]

$$(t_s[\mu] \to \langle t, t_s\rangle_r[\mu])_N = \inf_{d \in dom(\mathcal{A}_S)} \max(1 - t_s[\mu_\Pi], 1 - \Pi(=\mid d, t_s[\mathcal{A}_S]), \max_{t'} \min(N(=\mid d, t'_r[\mathcal{A}_S]), \langle t, t'\rangle_r[\mu_N])), \quad (6)$$

$$(t_s[\mu] \to \langle t, t_s\rangle_r[\mu])_\Pi = \inf_{d \in dom(\mathcal{A}_S)} \max(1 - t_s[\mu_N], 1 - N(=\mid d, t_s[\mathcal{A}_S]), \max_{t'} \min(\Pi(=\mid d, t'_r[\mathcal{A}_S]), \langle t, t'\rangle_r[\mu_\Pi])), \quad (7)$$

where $dom(\mathcal{A}_S) = dom(A_1) \times dom(A_2) \times \cdots \times dom(A_m)$ when $\mathcal{A}_S = \{A_1, A_2, \ldots, A_m\}$, and $\langle t, t'\rangle_r$ is a tuple with $\langle t, t'\rangle_r[\mathcal{A}_{R-S}] = t$ and $\langle t, t'\rangle_r[\mathcal{A}_S] = t'_r[\mathcal{A}_S]$ in relation r, and d, which is contained in $Dom(\mathcal{A}_S)$, is a sequence $\langle d[A_1], d[A_2], \ldots, d[A_m]\rangle$ of possible value $d[A_l]$ of attribute value $t_s[A_l]$ in relation s, and

$$N(=\mid d, t_s[\mathcal{A}_S]) = \min_{A_l \in \mathcal{A}_S} N(=\mid d[A_l], t_s[A_l]),$$

$$\Pi(=\mid d, t_s[\mathcal{A}_S]) = \min_{A_l \in \mathcal{A}_S} \Pi(=\mid d[A_l], t_s[A_l]).$$

According to the reference[15],

$$N(=\mid d[A_l], t_s[A_l]) = \inf_{u \in dom(A_l)} \max(\mu_=(d[A_l], u), 1 - \pi_{t_s[A_l]}(u)),$$

$$\Pi(=\mid d[A_l], t_s[A_l]) = \sup_{u \in dom(A_l)} \min(\mu_=(d[A_l], u), \pi_{t_s[A_l]}(u)),$$

where if $x = y$ then $\mu_=(x, y) = 1$, otherwise $\mu_=(x, y) = 0$.[§]

[‡]The formulation is based on possible values. The other formulation based on possible relations is proposed for relations with a single-valued membership attribute[7]. They derive the same results when relations have a single-valued membership attribute.

[§]In this case, a resemblance between entities is not considered. The resemblance usually exists in the real world. A resemblance relation is used for $\mu_=$ when the resemblance is built in databases. However, some pieces of information about resemblance degrees should be given users in query processing, otherwise misunderstanding of results may occur, as is pointed out by reference [10].

3.2 Flexible division operators

The formula mentioned so far is the expression that corresponds to the universal quantifier *for all.* Taking min for t_s in formulae (2) and (3) corresponds to using *for all.* Some researchers have tried to make division operators flexible; namely, to relax *for all* so that we can use a type of quantifiers *exists, at least k, most* and so on[2, 5, 8, 11]. One approach is to evaluate to what extent exceptions are allowed[5, 8, 11]. The other approach is to use linear interpolation between *for all* and *exists*[2]. These approaches are made under a restriction where usable quantifiers have some properties; namely, they have to be expressed by a non-decreasing membership function. However, all types of quantifiers cannot be expressed by a non-decreasing membership function. There are other types of quantifiers; *non-increasing* and *unimodal.* For example, a non-increasing type of quantifier *at most three* and unimodal type of quantifiers *about three* and *several,* which cannot be expressed by a non-decreasing membership function, are not available in the previous approaches. Moreover, the investigation has been made under a limited framework that a tuple has a single-valued membership attribute value and attribute values are precise in relations. As a matter of fact, many authors allow imprecise values as attribute values and impose fuzzy restrictions on relations[13–16]. In this case, what we can know is a pair of degrees in necessity and in possibility for compatibility of a tuple value with restrictions imposed on a relation. This is the reason why we make a division operator flexible in the possibility-based fuzzy relational model based on necessity and possibility measures, which is addressed in section 2.

A flexible division operator is much related with a quantified statement of the type "Q X G are H"[5, 25]. The statement means that Q elements of set X that are compatible with predicate G are also compatible with predicate H. This is the same as that the count (resp. the ratio) of which elements of set X that are compatible with G are also compatible with H is Q, when Q is an absolute quantifier (resp. a proportional quantifier). Thus, "Q X G are H" is equivalent to "X G are H is Q." To evaluate "Q X G are H" is to calculate compatibility degrees of a count or a ratio of "X G are H" with Q.

In division $r \div_Q s$, X G is relation s; G is a restriction that relation s has; H is to exist as a subtuple value in relation r; "X G are H" is that tuple values of relation s exist as a subtuple value accompanied by the same other subtuple value when a tuple value is divided into two subtuple values in relation r. For example, let us consider a query "*Select suppliers that sufficiently supply a few of expensive parts used in tool T.*" G is to be expensive and used in tool T; X G is a set of parts that are expensive and used in tool T; H is to be sufficiently supplied by a supplier; "X G is H" is that a set of parts that are expensive and used in tool T are sufficiently supplied by the same supplier; Q is $afew$. When fuzzy relational schemes of relation *supply* and of relation *part* are ($SUPPLIER$, $PART$, $QUANTITY$, (μ_N, μ_Π)) and ($TOOL$, $PART$, $PRICE$, (μ_N, μ_Π)), respectively, s is a derived relation from relation *part* where $TOOL$ is T and $PRICE$ is expensive; r is a derived relation from relation *supply* where $QUANTYTY$ is sufficient.

At the beginning, we focus on the case that an absolute quantifier is used. First, we have to obtain count C_t of elements composing a set of $(t_s[\mu] \rightarrow \langle t, t_s \rangle_r[\mu])/t_s[\mathcal{A}_S]$ with common subtuple value t for every tuple value $t_s[\mathcal{A}_S]$ in calculating $r \div_Q s$. However, we cannot exactly know the count, because we can obtain only to what degrees in necessity and in possibility an imprecise attribute value is compatible with fuzzy restrictions imposed on a relation. What we can know is count C_t^* from degrees of necessity and count C_{t*} from degrees of possibility.

Let Z_t be a set of $(t_s[\mu] \rightarrow \langle t, t_s \rangle_r[\mu])/t_s[\mathcal{A}_S]$. When relation s consists of k tuples,

$$Z_t = \{(\mu_{N,1}, \mu_{\Pi,1})/t_{s,1}[\mathcal{A}_S], (\mu_{N,2}, \mu_{\Pi,2})/t_{s,2}[\mathcal{A}_S], \cdots, \\ (\mu_{N,k}, \mu_{\Pi,k})/t_{s,k}[\mathcal{A}_S]\}, \qquad (8)$$

where $t_{s,i}$ is a tuple in relation s, $\mu_{N,i} = (t_{s,i}[\mu] \rightarrow \langle t, t_{s,i} \rangle_r[\mu])_N$,and $\mu_{\Pi,i} = (t_{s,i}[\mu] \rightarrow \langle t, t_{s,i} \rangle_r[\mu])_\Pi$. When $Z_{t,N}$ (resp. $Z_{t,\Pi}$) denotes a part of necessity (resp. possibility),

$$Z_{t,N} = \{\mu_{N,1}/t_{s,1}[\mathcal{A}_S], \mu_{N,2}/t_{s,2}[\mathcal{A}_S], \cdots, \mu_{N,k}/t_{s,k}[\mathcal{A}_S]\}, \qquad (9)$$

$$Z_{t,\Pi} = \{\mu_{\Pi,1}/t_{s,1}[\mathcal{A}_S], \mu_{\Pi,2}/t_{s,2}[\mathcal{A}_S], \cdots, \mu_{\Pi,k}/t_{s,k}[\mathcal{A}_S]\}. \qquad (10)$$

Counts of elements consisting of $Z_{t,N}$ and $Z_{t,\Pi}$ are C_{t*} and C_t^*, respectively.

$\mu_{N,i}/t_{s,i}[\mathcal{A}_S]$, which is obtained from necessity measures, means that $t_{s,i}[\mathcal{A}_S]$ exists at least to degree $\mu_{N,i}$, because a degree in necessity for an event is the complementation of a degree in possibility for its opposite event to 1. This leads to that membership function $\mu_{|Z_{t,N}|}$ of cardinality $|Z_{t,N}|$ is calculated by

$$\mu_{|Z_{t,N}|}(j) = \sup\{\alpha \,|\, |Z_{t,N,\alpha}| = j\}, \qquad (11)$$

where $Z_{t,N,\alpha} = \{u | \mu_{Z_{t,N}}(u) \geq \alpha\}$, called the α level set of $Z_{t,N}$. Note that $\mu_{|Z_{t,N}|}(j) = 0$ when there is no α with $|Z_{t,N,\alpha}| = j$. This matches what a degree from necessity measures means, as is shown in the following example.

Example

When $Z_{t,N} = \{1/t_{s,1}[\mathcal{A}_S], 1/t_{s,2}[\mathcal{A}_S], 0.8/t_{s,3}[\mathcal{A}_S], 0.8/t_{s,4}[\mathcal{A}_S], 0.3/t_{s,5}[\mathcal{A}_S]\}$ is given, by using formula (11) we obtain $|Z_{t,N}| = 1/2 + 0.8/4 + 0.3/5$. It is natural that $|Z_{t,N}|$ cannot take 0 or 1, since two elements have the maximum degree 1 of necessity in $Z_{t,N}$ which means the two elements certainly exists. Similarly, $|Z_{t,N}|$ is 4 to a degree 0.8 and cannot take 3.

On the other hand, $\mu_{\Pi,i}/t_{s,i}[\mathcal{A}_S]$, which is obtained from possibility measures, means that $t_{s,i}[\mathcal{A}_S]$ exists at most to degree $\mu_{\Pi,i}$ and may be nonexistent. This leads to that membership function $\mu_{|Z_{t,\Pi}|}$ of cardinality $|Z_{t,\Pi}|$ is calculated by

$$\mu_{|Z_{t,\Pi}|}(j) = \sup\{\alpha \| Z_{t,\Pi,\alpha}| \geq j\}, \tag{12}$$

where $Z_{t,\Pi,\alpha}$ is the α level set of $Z_{t,\Pi}$. Note that $\mu_{Z_{t,\Pi}}(0) = 1$ and $\mu_{Z_{t,\Pi}}(j) \geq \mu_{Z_{t,\Pi}}(j+1)$. This matches what a degree from possibility measures means, as is shown in the following example.

Example
When $Z_{t,\Pi} = \{1/t_{s,1}[\mathcal{A}_S], 1/t_{s,2}[\mathcal{A}_S], 0.8/t_{s,3}[\mathcal{A}_S], 0.8/t_{s,4}[\mathcal{A}_S], 0.6/t_{s,5}[\mathcal{A}_S]\}$ is given, by using formula (12) we obtain $|Z_{t,\Pi}| = 1/0 + 1/1 + 1/2 + 0.8/3 + 0.8/4 + 0.6/5$. Although two elements have the maximum degree 1 of possibility $Z_{t,\Pi}$, the maximum degree means that the two elements do not exist certainly, but only does at most. They may exist or not. Thus, we have the maximum degree of possibility that $|Z_{t,\Pi}|$ can take 0 or 1. Similarly, $|Z_{t,\Pi}|$ can take 3 and 4 to a degree 0.8.

We still have something to consider. We have to note to obtain a pair of degrees from necessity and possibility measures. For example, when we obtain $Z_t = \{(1,1)/t_{s,1}[\mathcal{A}_S], (1,1)/t_{s,2}[\mathcal{A}_S], (0.8,0.8)/t_{s,3}[\mathcal{A}_S], (0.8,0.8)/t_{s,4}[\mathcal{A}_S], (0.3,0.6)/t_{s,5}[\mathcal{A}_S]\}$,

$$\begin{aligned} Z_{t,N} &= \{1/t_{s,1}[\mathcal{A}_S], 1/t_{s,2}[\mathcal{A}_S], 0.8/t_{s,3}[\mathcal{A}_S], 0.8/t_{s,4}[\mathcal{A}_S], 0.3/t_{s,5}[\mathcal{A}_S]\}, \\ Z_{t,\Pi} &= \{1/t_{s,1}[\mathcal{A}_S], 1/t_{s,2}[\mathcal{A}_S], 0.8/t_{s,3}[\mathcal{A}_S], 0.8/t_{s,4}[\mathcal{A}_S], 0.6/t_{s,5}[\mathcal{A}_S]\}. \end{aligned}$$

As is shown in the previous two examples,

$$\begin{aligned} |Z_{t,N}| &= 1/2 + 0.8/4 + 0.3/5, \\ |Z_{t,\Pi}| &= 1/0 + 1/1 + 1/2 + 0.8/3 + 0.8/4 + 0.6/5. \end{aligned}$$

As a matter of fact, this $|Z_{t,\Pi}|$ is not suitable to count C_t^* for possibility. A degree from necessity measures is considered as a restriction to a degree from possibility measures; namely, the lower limit of the degree in possibility. In the above Z_t two elements have the maximum degree of necessity equal to 1; namely, the two elements exist certainly. Thus, the count has no possibility that it takes 0 or 1. Similarly, the count is 4, not 3 to a degree 0.8. This is not reflected in $|Z_{t,\Pi}|$. We cannot obtain a correct distribution from degrees in possibility without considering degrees in necessity. Thus, we use for membership function $\pi_{C_{t*}}$ (resp. $\pi_{C_t^*}$) of C_{t*} (resp. C_t^*) the following rules:

- $\forall i \;\; \pi_{C_{t*}}(j) = \mu_{|Z_{t,N}|}(j)$.
- $\pi_{C_t^*}(j) = \mu_{|Z_{t,\Pi}|}(j)$ except for the below two cases:
  if $\mu_{Z_{t,N}}(l) = 1$, then $\pi_{C_t^*}(j) = 0 \; for \; j < l$;
  if $\mu_{Z_{t,\Pi}}(j) = \mu_{Z_{t,N}}(j+1)$[¶], then $\pi_{C_t^*}(j) = 0$.

¶This condition means $\mu_{Z_{t,\Pi}}(j) = \mu_{Z_{t,\Pi}}(j+1) = \mu_{Z_{t,N}}(j+1)$ and $\mu_{Z_{t,N}}(j) = 0$, as is proved easily.

Under these rules, $C_{t*} = C_t^*$ holds if $Z_{t,N} = Z_{t,\Pi}$ which corresponds to the cases that the attribute values are precise or no fuzzy restrictions are imposed on relations even if attribute values are imprecise. This shows that the rules are reasonable.

Let degrees to which C_{t*} and C_t^* are compatible with a quantifier Q be denoted by $(N(Q|C_{t*}), \Pi(Q|C_{t*}))$ and $(N(Q|C_t^*), \Pi(Q|C_t^*))$, respectively. According to Prade & Testemale[15],

$$N(Q|C_{t*}) = \min_{j=0,k} \max(1 - \pi_{C_{t*}}(j), \mu_Q(j)), \quad (13)$$

$$\Pi(Q|C_{t*}) = \max_{j=0,k} \min(\pi_{C_{t*}}(j), \mu_Q(j)), \quad (14)$$

$$N(Q|C_t^*) = \min_{j=0,k} \max(1 - \pi_{C_t^*}(j), \mu_Q(j)), \quad (15)$$

$$\Pi(Q|C_t^*) = \max_{j=0,k} \min(\pi_{C_t^*}(j), \mu_Q(j)), \quad (16)$$

where μ_Q is a membership function of quantifier Q.
$N(Q|C_{t*}) \leq \Pi(Q|C_{t*})$ and $N(Q|C_t^*) \leq \Pi(Q|C_t^*)$ hold, because C_{t*} and C_t^* are normal. $N(Q|C_t^*) \leq N(Q|C_{t*})$ and $\Pi(Q|C_{t*}) \leq \Pi(Q|C_t^*)$ are valid, because $\mu_{C_{t*}}(j) \leq \mu_{C_t^*}(j)$ for $j = 0, \cdots, k$. Thus, the minimum compatibility degree is $N(Q|C_t^*)$, the maximum one $\Pi(Q|C_t^*)$. In addition, C_t^* is a possibility distribution, because it is basically obtained from degrees by possibility measures. Thus, it is reasonable to set $N(Q|C_t^*)$ as a degree in necessity and $\Pi(Q|C_t^*)$ as a degree in possibility for a membership value of t in $r \div_Q s$. Interestingly, formula $\Pi(Q|C_t^*)$ for possibility coincides with the formula using Sugeno integral when Q is a non-decreasing type of quantifiers[6].

Example
Let the following query be given: "*Select suppliers that sufficiently supply a few of expensive parts used in tool T.*" Relation r is obtained from a relation with fuzzy relational scheme $(SUPPLIER, PART, QUANTITY, (\mu_N, \mu_\Pi))$ under a restriction "*QUANTYTY is sufficient.*" Relation s is obtained from a relation with fuzzy relational scheme $(TOOL, PART, PRICE, (\mu_N, \mu_\Pi))$ under a restriction "*TOOL is T and PRICE is expensive.*" ! !

r

$SUPPLIER$	$PART$	μ
$1/a$	$1/v$	1,1
$1/a$	$1/v'$	1,1
$1/a$	$1/w$	1,1
$1/a$	$1/x$	0.8,0.8
$1/a$	$1/y$	0.8,0.8
$1/a$	$1/z + 0.7/z'$	0.6,0.6
$1/b$	$1/w$	1,1
$1/b$	$1/x$	1,1

s

$PART$	μ
$1/v + 0.8/v'$	0.9,1
$1/w$	0.7,1
$1/x$	0.3,1
$1/y$	1,1
$1/z$	1,1

By using formulae (6) and (7) for $1/a$ of $SUPPLIER$, we obtain $Z_{1/a}$ expressed by formula (8): $Z_{1/a} = \{(1,1)/(1/v+0.8/v'), (1,1)/(1/w), (0.8,0.8)/(1/x), (0.8,0.8)/(1/y), (0.3,0.6)/(1/z)\}$. From this $Z_{1/a}$,

$$
\begin{aligned}
Z_{1/a,N} &= \{1/(1/v+0.8/v'), 1/(1/w), 0.8/(1/x), 0.8/(1/y), 0.3/(1/z)\},\\
Z_{1/a,\Pi} &= \{1/(1/v+0.8/v'), 1/(1/w), 0.8/(1/x), 0.8/(1/y), 0.6/(1/z)\}.
\end{aligned}
$$

Thus,

$$
\begin{aligned}
|Z_{1/a,N}| &= 1/2 + 0.8/4 + 0.3/5,\\
|Z_{1/a,\Pi}| &= 1/0 + 1/1 + 1/2 + 0.8/3 + 0.8/4 + 0.6/5.
\end{aligned}
$$

Using the above rules,

$$
\begin{aligned}
C_{1/a*} &= 1/2 + 0.8/4 + 0.3/5,\\
C^*_{1/a} &= 1/2 + 0.8/4 + 0.6/5.
\end{aligned}
$$

Suppose that $afew$ is expressed by $0.3/1 + 1/2 + 1/3 + 0.3/4$. We obtain $N(Q|C^*_{1/a}) = 0.3$ and $\Pi(Q|C^*_{1/a}) = 1$. By doing similar calculations for $1/b$ of $SUPPLIER$, the following derived relation is obtained:

$r \div_{afew} s$

$SUPPLIER$	μ
$1/a$	0.3,1
$1/b$	1,1

When Q is a proportional quantifier, ratio R_t is used and a set of rational numbers $\{0, \frac{1}{k}, \frac{2}{k}, \cdots, \frac{k-1}{k}, 1\}$ is in place of a set of integers $\{0, 1, 2, \cdots, k-1, k\}$.

$$
N(Q|R^*_t) = \min_{j=0,k} \max(1 - \pi_{R^*_t}(\frac{j}{k}), \mu_Q(\frac{j}{k})), \tag{17}
$$

$$
\Pi(Q|R^*_t) = \max_{j=0,k} \min(\pi_{R^*_t}(\frac{j}{k}), \mu_Q(\frac{j}{k})). \tag{18}
$$

Dubois and Prade[14, 26] obtain a possibility distribution of count C by using min operation of a possibility distribution of at least k obtained from degrees in possibility and that of at most k from degrees in necessity. However, their approach does not guarantee the obtained possibility distribution to be normal. Prade[27] obtains a normal possibility distribution by using a necessity measure different from the ordinary necessity measure, which is derived from para-consistent complementation. We cannot use a particular necessity measure only to a division operator that is one of eight operators composing relational algebra.

4 Concluding remarks

We have shown a flexible division operator in a possibility-based fuzzy relational model based on necessity and possibility measures. A membership attribute value of t in $r \div_Q s$ is a degree to which a possibility distribution obtained from a set of $(t_s[\mu] \rightarrow \langle t, t_s\rangle_r[\mu])/t_s[\mathcal{A}_S]$ is compatible with quantifier Q. The approach has no restrictions to types of quantifiers, although the methods proposed so far do so. This is very significant, because users can use the flexible division operator without paying attention to what type quantifiers belong to.

Acknowledgments
The author wishes to thank the anonymous reviewers for their comments, which were valuable in improving the quality of the final version.

References

1. Codd, E. F. [1972] Relational Completeness of Data Base Sublanguage, in: Data Base Systems, R. Rustin, ed., Prentice-Hall, Englewood Cliffs, New Jersey, pp. 33-64.

2. Cubero, J. C., Medina, J. M., Pons, O., and Vila, M. A. [1994] The Generalized Selection: An Alternative Way for the Quotient Operations in Fuzzy Relational Databases, in: Proceedings of the 5th International Conference on Information Processing and Management of Uncertainty in Knowledge-Based System (IPMU '94), Paris, France, July 4-8, 1994, pp. 23-30.

3. Mouaddib, N. [1994] Fuzzy Identification in Fuzzy Databases. The Nuanced Relational Division, International Journal of Intelligent Systems, **9**, 461-473.

4. Umano, M. and Fukami, S. [1994] Fuzzy Relational Algebra for Possibility-Distribution-Fuzzy-Relational Model of Fuzzy Data, Journal of Intelligent Information Systems, **3**, 7-27.

5. Yager, R. R. [1994] Interpreting Linguistically Qualified Propositions, International Journal Intelligent Systems, **9**, 541-569.

6. Bosc, P. and Liétard, L. [1995] On the Comparison of the Sugeno and the Choquet Integrals for the Evaluation of Quantified Statements, in: Proceedings of the 3rd European Congress on Intelligent Techniques and Soft Computing (EUFIT '95), Aachen, Germany, pp. 709-716.

7. Bosc, P. and Liétard, L. [1995]On the division of relations with imprecise information, in: Fuzzy division for regular relational databases.

in: Foundations and Applications of Possibility Theory (Proceedings of FART '95, Gent, Belgium, Dec. 13-15, 1995.), G. de Gooman, Da Ruan, and E. E. Kerre, eds., World Scientific, Singapore, pp. 287-294.

8. Dubois, D., and Prade, H. [1996] Semantics of Quotient Operators in Fuzzy Relational Databases, Fuzzy Sets and Systems, **78**, 89-93.

9. Bosc, P., Dubois, D., Pivert, O., and Prade, H. [1997] Flexible queries in relational databases – The example of the division operator, Theoretical Computer Science, **171**, 281-302.

10. Bosc, P. and Liétard, L. [1997] Different Relaxations of the Relational Division, in: Proceedings of the Seventh International Fuzzy Systems Association World Congress (IFSA '97), Vol. II, Prague, Czech Republic, June 25-29, 1997, pp. 180-185.

11. Dubois, D., Nakata, M., and Prade, H. [2000] Extended division for flexible queries in relational databases, in Knowledge Management in Fuzzy Databases, Eds., O. Pons, M. A. Vila., and J. Kacprzyk, Physica Verlag, pp. 105-121.

12. Nakata, M. [2000]Formulating Division Operators in fuzzy relational databases, in Knowledge Management in Fuzzy Databases, Eds., O. Pons, M. A. Vila., and J. Kacprzyk, Physica Verlag, pp. 144-156.

13. Umano, M. [1982] FREEDOM-O: A Fuzzy Database System, in: Fuzzy Information and Decision Processes, M. M. Gupta and E. Sanchez, eds., North-Holland, Amsterdam, pp. 339-347.

14. Prade, H. [1984] Lipski's Approach to Incomplete Information Data Bases Restated and Generalized in the Setting of Zadeh's Possibility Theory. Information Systems **9**:1, 27-42.

15. Prade, H. and Testemale, C. [1984] Generalizing Database Relational Algebra for the Treatment of Incomplete or Uncertain Information and Vague Queries, Information Sciences, **34**, 115-143.

16. Zemankova, M. and Kandel, A. [1984] Fuzzy Relational Databases–A Key to Expert Systems, Verlag TÜV Rheinland, Cologne.

17. Raju, K. V. S. V. N. and Majumdar, A. K. [1988] Fuzzy Functional Dependencies and Lossless Join Decomposition of Fuzzy Relational Database Systems, ACM Transactions on Database Systems, **13**(2), 129-166.

18. Zadeh, L. A. [1965] Fuzzy Sets, Information and Control, **12**, 338-353.

19. Zadeh, L. A. [1978] Fuzzy Sets as a Basis for a Theory of Possibility, Fuzzy Sets and Systems, **1**, 3-28.

20. Dubois, D. and Prade, H. with the Collaboration of H. Farreny, R. Martin-Clouaire and C. Testemale, [1988] Possibility Theory: An Approach to Computerized Processing of Uncertainty, Plenum Publishing Co.

21. Baldwin, J. F. [1983] A Fuzzy Relational Inference Language for Expert Systems, in: Proceedings of the 13th IEEE International Symposium on Multiple-Valued Logic, Kyoto, Japan, IEEE Computer Society Press, pp. 416-421.

22. Umano, M. [1983] Retrieval from Fuzzy Database by Fuzzy Relational Algebra, in: Proceedings of IFAC Symposium, Fuzzy Information, Knowledge Representation and Decision Analysis, E. Sanchez, ed., Marseille, France, July 19-21, 1983, Pergamon Press, pp. 1-6.

23. Li, D. and Liu, D. [1990] A Fuzzy Prolog Database System, Research Studies Press.

24. Ullman, J. [1982] Principles of Database Systems, 2nd edition, Computer Science Press.

25. Bosc, P., Liétard, L., and Prade, H. [1996] On fuzzy queries involving fuzzy quantifiers, in: Proceedings of the ECAI Workshop on Uncertainty in Information Systems, Budapest, Hungary, Aug. 11-16, 1996, pp. 6-10. Extended version "An ordinal approach to the processing of fuzzy queries with flexible quantifiers" in: Applications of Uncertainty Formalisms (A. Hunter and S. Parsons, eds.), Lecture Notes in Computer Science Vol. 1455, Springer 1998.

26. Dubois, D. and Prade, H. [1985] Fuzzy Cardinality and the Modeling of Imprecise Qualification, Fuzzy Sets and Systems **16** 199-230.

27. Prade, H. [1990] A Two-Layer Fuzzy Pattern Matching Procedure for Evaluation of Conditions Involving Vague Quantifiers, Journal of Intelligent and Robotic Systems, **3**, 93-101.

Part II.

Data Mining Based on Fuzzy Sets

Fuzzy Data Mining: Discovery of Fuzzy Generalized Association Rules[†]

Guoqing Chen[1], Qiang Wei[1], Etienne E. Kerre[2]

[1] Division of Management Science & Engineering, School of Economics & Management, Tsinghua University, Beijing 100084, China
[2] Department of Applied Mathematics, University of Gent, Krijgslaan 281/S9, 9000 Gent, Belgium

Abstract. Data mining is a key step of knowledge discovery in databases. Classically, mining generalized association rules is to discover the relationships between data attributes upon all levels of presumed exact taxonomic structures. In many real-world applications, however, the taxonomic structures may not be crisp but fuzzy. This paper focuses on the issue of mining generalized association rules with fuzzy taxonomic structures. First, fuzzy extensions are made to the notions of the degree of support, the degree of confidence, and the R-interest measure. The computation of these degrees takes into account the fact that there may exist a partial belonging between any two itemsets in the taxonomy concerned. Then, the classical Srikant and Agrawal's algorithm (including the Apriori algorithm and the Fast algorithm) is extended to allow discovering the relationships between data attributes upon all levels of fuzzy taxonomic structures. In this way, both crisp and fuzzy association rules can be discovered. Finally, the extended algorithm is run on the synthetic data with up to 10^6 transactions. It reveals that the extended algorithm is at the same level of computational complexity in the number of the transactions as that of the classical algorithm.

Keywords. Generalized association rules, Degree of support, Degree of confidence, Fuzzy taxonomic structure

1. Introduction

Data mining is a key step of knowledge discovery in large databases. One of the important issues in the field is to efficiently discover the relationships among data items in forms of association rules that are of interest to decision-makers. In 1993, Agrawal [1] proposed an algorithm for mining association rules that represent the relationships between basic data items (e.g., items from original sales records). An example of such rules is "the customers who bought apples might also turn to buy pork". More concretely, if of all the customers, 20% bought both apples and pork, and of the customers who bought apples, 80% also bought pork, then the rule, represented in form of Apple $\Rightarrow$ Pork, is regarded to be with the degree of

[†] Supported by the National Outstanding Youth Foundation of China.

confidence 80% and the degree of support 20%. Given the rule, the manager, or decision-maker of the super market, may consider to place pork near apples in order to improve the sales. In recent years, various efforts have been made to improve or extend the algorithm, e.g., [2,3, 5-7, 9-10, 12-14]. In Srikant and Agrawal [13], the algorithm is extended to allow the discovery of the so-called generalized association rules that represent the relationships between basic data items, as well as between data items at all levels of related taxonomic structures. In most cases, taxonomies (is-a hierarchies) over the items are available [13]. An example of taxonomic structures is shown in Figure 1.

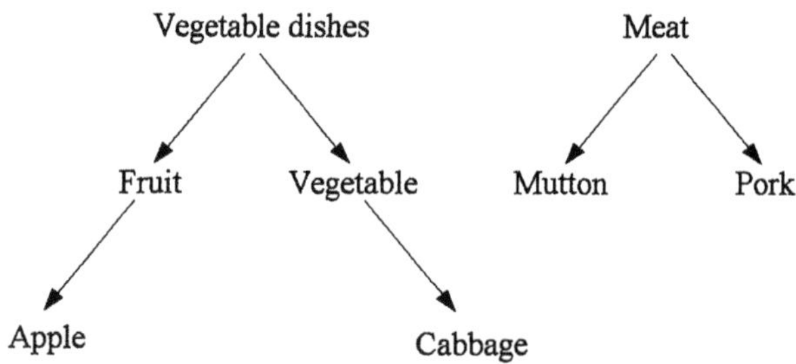

Figure 1 Example of taxonomic structures

With such taxonomies, the generalized association rules like Fruit $\Rightarrow$ Meat for Figure 1 are often meaningful: the rules with respect to lower levels of taxonomic structures (e.g., Cabbage $\Rightarrow$ Pork) may not be "significant" enough to be mined according to the mining criteria, while the rules with respect to related higher levels of the structures may be discovered due to the existence of strong associations. In fact, the rules at high levels often reflect more abstract and meaningful business semantics. Notably, the computation of the degree of support, denoted hereafter as Dsupport, and the degree of confidence, denoted hereafter as Dconfidence, plays an important role in the algorithm [13]. Specifically, Dsupport and Dconfidence of the generalized association rule X$\Rightarrow$Y are defined as follows:

$$Dsupport\,(X \Rightarrow Y) = \|X \cup Y\| / |T|$$
$$Dconfidence\,(X \Rightarrow Y) = \|X \cup Y\| / \|X\|$$

Where X and Y are itemsets with $X \cap Y = \emptyset$, T is the set of all the transactions contained in the database concerned, $\|X\|$ is the number of the transactions in T that contain X, $\|X \cup Y\|$ is the number of the transactions in T that contain X and Y, and $|T|$ is the number of the transactions contained in T. Mining generalized association rules X$\Rightarrow$Y, if any, in a database is to find whether the transactions in the database satisfy the pre-specified thresholds, min-support and min-confidence, for Dsupport and Dconfidence respectively. Usually, the rules discovered in this way may need to be further filtered, for instance, to eliminate redundant and inconsistent rules using the R-interest measure, which will be discussed in Section 2.

However, in many real world applications, the related taxonomic structures may not be necessarily crisp, rather, certain fuzzy taxonomic structures reflecting partial belonging of one item to another may pertain. For example, Tomato may be regarded as being both Fruit and Vegetable, but to different degrees. An example of a fuzzy taxonomic structure is shown in Figure 2. Here, a sub-item belongs to its super-item with a certain degree. Apparently, in such a fuzzy context, the computation of Dsupport and Dconfidence shown above can hardly be applied, but needs to be extended accordingly.

Furthermore, the algorithm [13] used in discovering the generalized association rules needs to be extended as well. This involves the incorporation of fuzziness, for instance, for the generation of frequent itemsets (e.g., Apriori algorithm) and for the generation of the rules from the frequent itemsets (e.g., Fast algorithm), as well as for the generation of the extended transaction set T'.

In section 2, the taxonomic structures are extended to allow partial belongings between itemsets. In the mean time, the computation of fuzziness-involved Dsupport, Dconfidence and R-interest is discussed. Section 3 explores the extension to the classical algorithm based on the extended notions of Dsupport, Dconfidence and R-interest discussed in section 2. In section 4, the extended algorithm is run on the synthetic data to help reveal certain aspects of its performance as compared with that of the classical algorithm. Finally, section 5 will conclude the current work and highlight some of the ongoing and future studies.

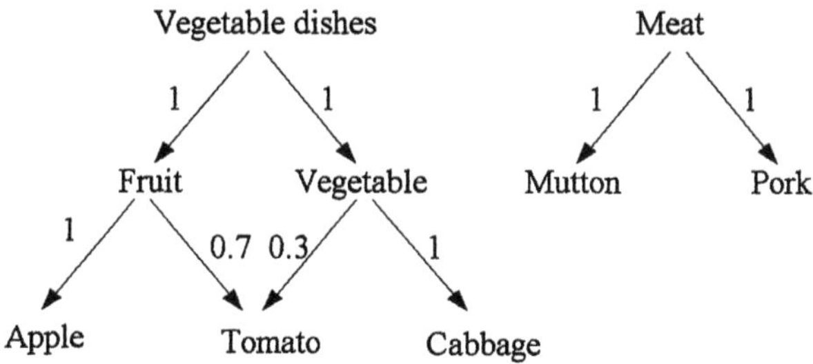

Figure 2 Example of fuzzy taxonomic structures

2. Fuzzy Taxonomic Structures

2.1 Fuzzy Extension to Crisp Taxonomic Structures

A crisp taxonomic structure assumes that the child item belongs to its ancestor with degree 1. But in a fuzzy taxonomy, this assumption is no longer true. Different degrees may pertain across all nodes (itemsets) of the structure.

Let I = $\{i_1, i_2, \ldots, i_m\}$ be a set of literals, called items. Let FG be a directed a-cyclic graph (DAG) on the literals [15]. An edge in FG represents a fuzzy *is-a* relationship, which means along with each edge, there exists a partial degree μ with which the child-node on this edge belongs to its parent-node on this edge, where $0 \le \mu \le 1$. If there is an edge in FG from *p* to *c*, *p* is called a parent of *c* and *c* a child of *p* (*p* represents a generalization of *c*.). The fuzzy taxonomic structure is defined as a DAG rather than a forest to allow for multiple taxonomies.

We call x^ an *ancestor* of x (and x a descendant of x^) if there is a directed path (a series of edges) from x^ to x in FG. Note that a node is not an ancestor of itself, since the graph is a-cyclic.

Let T be a set of all transactions, I be a set of all items, and t be a transaction in T such that $t \subseteq I$. Then, we say that a transaction t supports an item $x \in I$ with degree 1 if x is in t, or with degree μ , if x is an ancestor of some item y in t such that y belongs to x in a degree μ. We say that a transaction t supports $X \subseteq I$ with degree β:

$$\beta = \min_{x \in X}(\mu)$$

where μ is the degree to which x (in X) is supported by t, $0 \le \mu \le 1$.

In analogue to the crisp case, given a transaction set T, there may exist a fuzzy taxonomic structure FG as shown in Figure 3. In general, the degrees in the fuzzy taxonomic structures may be user-dependant or context-dependant.

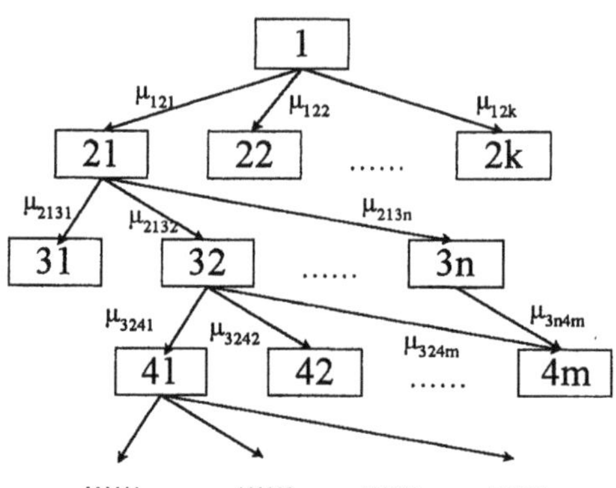

Figure 3 A fuzzy taxonomic structure

In Figure 3, every child-node x belongs to its parent-node y with degree μ_{yx}, $0 \le \mu_{yx} \le 1$. The leaf-nodes of the structure are attribute values of the transaction records. Every non-leaf-node is referred to as an attribute-node, which is regarded as a set whose elements are the leaf-nodes with respective membership degrees. Sometimes for the purposes of convenience, each leaf-node is regarded as a set that contains merely the attribute value itself. In this case, the attribute value belongs to the leaf-node with a degree of 1. As the fuzzy taxonomic structure represents the partial degrees of the edges, the degrees between leaf-nodes and attribute-nodes in FG need to be derived. This could be done based upon the notions of subclass, superclass and inheritance, which have been discussed in [4]. Specifically,

$$\mu_{xy} = \underset{\forall l:x\to y}{\oplus} \left(\underset{\forall e on l}{\otimes} \mu_{le} \right) \tag{1}$$

where l: $x \to y$ is one of the accesses (paths) of attributes x and y, e *on* l is one of the edges on access l, μ_{le} is the degree on the edge e on l. If there is no access between x and y, $\mu_{xy} = 0$. Notably, what specific forms of the operators to use for $\oplus$ and $\otimes$ depends on the context of the problems at hand. Merely for illustrative purposes, in this paper, ***max*** is used for $\oplus$ and ***min*** for $\otimes$.

2.2 Determining the Degree of Support and the Degree of Confidence

Now consider the computation of the degree of support in such a fuzzy taxonomic structure case. If a is an attribute value in a certain transaction $t \in T$, T is the transaction set, and x is an attribute in certain itemset X, then the degree μ_{xa} with which a belongs to x can be obtained according to formula (1). Thus, μ_{xa} may be viewed as the degree that the transaction $\{a\}$ supports x. Further, the degree that t supports X can be obtained as follows:

$$\mu_{tX} = Support_{tX} = \min_{x \in X} (\max_{a \in t} (\mu_{xa})) \tag{2}$$

In this way, the degree that a transaction t in T supports a certain itemset X is computed. Moreover, in terms of how many transactions in T support X, the $\Sigma count$ operator [4] is used to sum up all the degrees that are associated with the transactions in T:

$$Dsupport(X) = \sum_{t \in T} count\,(Support_{tX}) \,/\, |T| = \sum_{t \in T} count\,(\mu_{tX}) \,/\, |T| \tag{3}$$

Hence, for a generalized association rule X⇒Y, let X∪Y = Z ⊆ I, then Dsupport (X⇒Y) can be obtained as follows:

$$Dsupport(X \Rightarrow Y) = \sum_{t \in T} count\,(\mu_{tZ}) \,/\, |T| \tag{4}$$

In an analogous manner, *Dconfidence*($X \Rightarrow Y$) can be computed as follows:

$$Dconfidence(X \Rightarrow Y) = Dsupport(X \Rightarrow Y) \,/\, Dsupport(X) = \sum_{t \in T} count(\mu_{tZ}) \,/\, \sum_{t \in T} count(\mu_{tX}) \quad (5)$$

2.3 Filtering the Redundant Rules with R-interest

Based on (4) and (5) described above, all the rules with Dsupport and Dconfidence more than the pre-specified minimum degree of support and minimum degree of confidence can be obtained. But actually, there are still some "redundant" or "useless" rules. For example, consider the rule Fruit ⇒ Pork (20% Dsupport, 80% Dconfidence). If "Fruit" is a parent of "Tomato", and there are 100 transactions containing "Fruit" and 50 transactions containing "Tomato", then we may have a perception of 35 (50×0.7) transactions containing "Fruit" according to the taxonomic structures in Figure 2. We would expect the rule Tomato ⇒ Pork to have 7% (20%×35/100) Dsupport and 80% Dconfidence. If the actual Dsupport and Dconfidence for rule Tomato ⇒ Pork are really around 7% and 80% respectively, the rule can be considered redundant since it does not convey any additional information and is less general than the first rule (Fruit ⇒ Pork).

Thus, the concept of R-interest [13] can be extended based on the notion of Dsupport. Like in the classical case, the extended R-interest measure is a way used to prune out those "redundant" rules. Briefly speaking, the rules of interest, according to R-interest, are those rules whose degrees of support are more than R times the expected degrees of support or whose degrees of confidence are more than R times the expected degrees of confidence.

Consider a rule $X \Rightarrow Y$, where $X=\{x_1, x_2, \ldots, x_m\}$ and $Y=\{y_1, y_2, \ldots, y_n\}$. $X^\wedge$ and $Y^\wedge$ are called the ancestors of X and Y respectively, if $X^\wedge=\{x^\wedge_1, x^\wedge_2, \ldots, x^\wedge_m\}$ where $x^\wedge_i$ is an ancestor of x_i, $1 \le i \le m$, and $Y=\{y^\wedge_1, y^\wedge_2, \ldots, y^\wedge_n\}$, where $y^\wedge_j$ is an ancestor of y_j, $1 \le j \le n$. Then the rules $X^\wedge \Rightarrow Y$, $X^\wedge \Rightarrow Y^\wedge$ and $X \Rightarrow Y^\wedge$ are called the ancestors of the rule $X \Rightarrow Y$. Let $Dsupport_{E(X^\wedge \Rightarrow Y^\wedge)}(X \Rightarrow Y)$ denote the "expected" value of the degree of support of rule $X \Rightarrow Y$ and $Dconfidence_{E(X^\wedge \Rightarrow Y^\wedge)}(X \Rightarrow Y)$ denote the "expected" value of the degree of confidence, then with fuzzy taxonomic structures, we have

$$\begin{aligned} & Dsupport_{E(X^\wedge \Rightarrow Y^\wedge)}(X \Rightarrow Y) \\ & = \sum count(\mu_{t\{x_1\}}) \Big/ \sum count(\mu_{t\{x^\wedge_1\}}) \times \ldots \times \\ & \sum count(\mu_{t\{x_m\}}) \Big/ \sum count(\mu_{t\{x^\wedge_m\}}) \times \\ & \sum count(\mu_{t\{y_1\}}) \Big/ \sum count(\mu_{t\{y^\wedge_1\}}) \times \ldots \times \\ & \sum count(\mu_{t\{y_n\}}) \Big/ \sum count(\mu_{t\{y^\wedge_n\}}) \times \\ & Dsupport(X^\wedge \Rightarrow Y^\wedge) \end{aligned} \quad (6)$$

and

$$Dconfidence_{E(X^\wedge \Rightarrow Y^\wedge)}(X \Rightarrow Y)$$
$$= \sum count(\mu_{t\{y_1\}}) / \sum count(\mu_{t\{y^\wedge_1\}}) \times \ldots \times$$
$$\sum count(\mu_{t\{y_n\}}) / \sum count(\mu_{t\{y^\wedge_n\}}) \times \quad (7)$$
$$Dconfidence(X^\wedge \Rightarrow Y^\wedge)$$

According to (6) and (7), the expected values of Dsupport and Dconfidence of each rule could be obtained, which may be used to determine whether the rule is "interesting" or not.

Notably, in the case of crisp taxonomic structures, $\sum count(\mu_{t\{xi\}})$ and $\sum count(\mu_{t\{yi\}})$ degenerate to $\|\{x_i\}\|$ and $\|\{y_i\}\|$ respectively. Then (6) and (7) are the same as those given by Srikant and Agrawal [13].

2.4 An Example

Suppose that a supermarket maintains a database for the goods that customers have purchased as shown in Table 1.

Transaction #	Things Bought
#100	Apple
#200	Tomato, Mutton
#300	Cabbage, Mutton
#400	Tomato, Pork
#500	Pork
#600	Cabbage, Pork

Table 1 Transactions in a supermarket database

Leaf-nodes	The degrees of the ancestors and its own
Apple	1/Apple, 1/Fruit, 1/Vegetable dishes
Tomato	1/Tomato, 0.3/Vegetable, 0.7/Fruit, 0.7/Vegetable dishes
Cabbage	1/Cabbage, 1/Vegetable, 1/Vegetable dishes
Pork	1/Pork, 1/Meat
Mutton	1/ Mutton, 1/Meat

Table 2 Leaf-nodes and their ancestor's degrees

The min-support threshold is 30%, min-confidence is 60%, and R-interest is 1.2. Here, we should emphasis that these thresholds are context-dependant, which should be defined according to the concrete situation. It is assumed that the underlying taxonomic structures are fuzzy and as shown in Figure 1. Then, according to formula (1) we have Table 2 for those leaf-nodes and their ancestor's degrees. For instance, in Table 2, μ(Tomato$\in$Vegetable dishes) = max(min(1,0.7), min(1, 0.3)) = 0.7.

Furthermore, according to the formula (3) for the Σ*count* values, all the frequent itemsets are listed in Table 3 along with their corresponding Σ*count* values. Here, by a frequent itemset we mean the itemset whose Σ*count* value is more than min-support × |T|. In fact, in generating the frequent itemsets, we first compute all the candidate itemsets (whose Σ*count* values do not need to exceed min-support), from which the frequent itemsets are obtained by filtering with min-support.

Frequent Itemsets	Σ*count* values
{Cabbage}	2
{Tomato}	2
{Pork}	3
{Mutton}	2
{Fruit}	2.4
{Vegetable}	2.6
{Vegetable dishes}	4.4
{Meat}	5
{Cabbage, Meat}	2
{Tomato, Meat}	2
{Vegetable, Meat}	2.6
{Vegetable dishes, Meat}	3.4

Table 3 Σ*count* values for frequent itemset

In Table 3, the *(count* value for the itemset {Vegetable, Meat}, for example, is calculated as:

$$min(0.3, 1) + min(1, 1) + min(0.3, 1) + min(1, 1) = 2.6$$

Based on these Σ*count* values for all the frequent itemsets, the degrees of support for all candidate rules can be computed. Table 4 lists those rules discovered, which satisfy the given thresholds 30%, 60%, 1.2 for the degree of support, the degree of confidence, and R-interest, respectively. For instance, *Dsupport*(Vegetable dishes ⇒ Meat) = 3.4/6 = 57%, and *Dconfidence*(Vegetable dishes ⇒ Meat) = 3.4/4.4 = 77%.

Here, it should be mentioned that only frequent itemsets are used to generate association rules. This can be proven as follows:

First, according to (2),

$$\mu_{tX} = \min_{x\in X}(\max_{a\in t}(\mu_{xa})) \geq \min(\min_{x\in X}(\max_{a\in t}(\mu_{xa})), \min_{x\in Y}(\max_{a\in t}(\mu_{xa})))$$

$$= \min_{x\in X\cup Y}(\max_{a\in t}(\mu_{xa})) = \mu_{tX\cup Y}$$

Then,

$$Dsupport(X) = \sum_{t\in T} count\,(\mu_{tX}) / |T|$$

$$\geq \sum_{t\in T} count\,(\mu_{tX\cup Y}) / |T| = Dsupport(X\cup Y) = Dsupport(X\Rightarrow Y)$$

Second, given itemsets X and X∪Y and min-support, 1) if X∪Y is not a frequent itemset, then according to (4), Dsupport(X⇒Y) = Dsupport(X∪Y) < min-support. 2) if X is not a frequent itemset, which means that Dsupport(X) < min-support, then Dsupport(X⇒Y) ≤ Dsupport(X) < min-support.

Thus, association rules can only be generated from frequent itemsets. We can simply omit the non-frequent itemsets, which help improve the efficiency.

Similarly to its crisp counterpart, this approach generates the rules based upon given thresholds, namely min-support and min-confidence, which are usually set by domain experts or users according to the real application semantics at hand. In other words, these thresholds play the similar roles in this approach to that of the classical approach.

Next, we will examine rules with R-interest measure. It is worth mentioning that the rule Cabbage⇒Meat is filtered out, though with *Dsupport*(Cabbage ⇒ Meat) = 2/6 = 33% > 30% and *Dconfidence*(Cabbage ⇒ Meat) = 2/2 = 100% > 60%. This is done according to the R-interest measure:

$Dsupport_{E(\text{Vegetable}\Rightarrow\text{Meat})}(\text{Cabbage}\Rightarrow\text{Meat})$
$= \sum count(\mu_{t\{\text{Cabbage}\}}) / \sum count(\mu_{t\{\text{Vegetable}\}})\times$
$\sum count(\mu_{t\{\text{Meat}\}}) / \sum count(\mu_{t\{\text{Meat}\}})\times$
$Dsupport(\text{Vegetable}\Rightarrow\text{Meat})$
$= 2/2.6\times 5/5\times 2.6/6$
$= 33\%$

$Dconfidence_{E(\text{Vegetable}\Rightarrow\text{Meat})}(\text{Cabbage}\Rightarrow\text{Meat})$
$= \sum count(\mu_{t\{\text{Meat}\}}) / \sum count(\mu_{t\{\text{Meat}\}})\times$
$Dconfidence(\text{Vegetable}\Rightarrow\text{Meat})$
$= 5/5\times 100\%$
$= 100\%$

and Dsupport(Cabbage⇒Meat) / $\text{Dsupport}_{E(\text{Vegetable}\Rightarrow\text{Meat})}$ (Cabbage⇒Meat) = 33% / 33% = 1.0 < 1.2, and Dconfidence(Cabbage⇒Meat) / $\text{Dconfidence}_{E(\text{Vegetable}\Rightarrow\text{Meat})}$ (Cabbage⇒Meat) = 100% / 100% = 1.0 < 1.2, which means that this rule is regarded redundant with respect to the existing rule "Vegetable⇒Meat" in Table 4.

Interesting Rules	Dsupport	Dconfidence
Vegetable⇒Meat	43%	100%
Vegetable dishes⇒Meat	57%	77%
Meat⇒Vegetable dishes	57%	68%

Table 4 The discovered rules of interest

3. Mining Fuzzy Generalized Association Rules

The task of discovering generalized association rules with fuzzy taxonomic structures can be decomposed into four parts:

1. Determining the membership degree that each leaf attribute belongs to each of its ancestors.
2. Based on the membership degrees derived in part 1, finding all itemsets whose $\Sigma count$ values are greater than min-support $\times$ $|T|$. These itemsets are called frequent itemsets.
3. Using the frequent itemsets to generate the rules whose degrees of confidence are greater than the user-specified min-confidence.
4. Pruning all the uninteresting rules.

As mentioned previously, since the nodes of fuzzy taxonomies can be viewed generally as fuzzy sets (or linguistic labels) on the domains of leaf nodes, mining association rules across all levels of the nodes in the taxonomic structures means the discovery of fuzzy generalized association rules. Apparently, crisp generalized association rules are special cases.

3.1. The Extended Algorithm

There are a number of procedures or sub-algorithms involved in mining generalized association rules. Therefore, the extended algorithm proposed in this section is a collection of several sub-algorithms that perform respective functions. We will discuss those sub-algorithms in which fuzziness is incorporated.

First, recall the Srikant and Agrawal approach [13] in that all the ancestors of each leaf item in taxonomic structures are added into the transaction set T in order to form a so-called extended transaction set T'. In the case of fuzzy taxonomic structures, T' is generated by not only adding to T all the ancestors of each leaf item in fuzzy taxonomic structures, but also the degrees that the ancestors are supported by the transactions in T. This can be done by first determining the degrees that the leaf item belongs to its ancestors according to formula (1). Concretely, we have the following sub-algorithm Degree.

Sub-algorithm Degree:

```
forall leaf nodes LNi ∈ Taxonomy do
    forall interior nodes INj ∈ Taxonomy do
        μ(LNi, INj) = max_{∀l: INj→LNi}(min_{∀e on l}(μ_le)
        insert into Degree,
        values LNi, INj, μ(LNi, INj)
    endfor
endfor
```

As we can see, the codes given above are just super-code. Various existing algorithms may be used in implementing the ideas, such as the Dijkstra algorithm, Floyd algorithm, or matrix-product algorithm [11]. For instance, matrix-product algorithm is easier to understand while Floyd algorithm is more efficient.

Consequently, based upon the degrees in Degree, the extended transaction set T' can be generated by computing, for each transaction, the degree that the transaction supports the itemset concerned according to formula (2) and adding all such degrees into the transaction set T. Concretely, we have sub-algorithm Extended Transaction Set T' as follows:

Sub-algorithm Extended Transaction Set T':

```
forall t∈T do
    insert into T',
    values all the elements ∈t with degree 1,
            all the ancestor of elements ∈t with the degrees of support
    from t, Degree
endfor
```

Once the extended transaction set T' is generated, the next step is to generate the candidate itemsets. Hereby, the extension to the well-known Apriori algorithm [1, 13] is considered. Let C_k be the set of all candidate k-itemsets (potentially frequent itemsets) and L_k be the set of all frequent k-itemsets, where a k-itemset is an itemset that consists of k items. The major difference of the extended Apriori algorithm from the classical one is that, for all k, C_k and L_k are associated with their respective Σcount values as those shown in section 2. Concretely, the sub-algorithm Extended Apriori is given as follows:

Sub-algorithm Extended Apriori:

```
L_1 = {frequent 1-itemsets}
for {k = 2; L_{k-1} ≠ ∅; k++} do
    C_k = Apriori-Gen(L_{k-1});   // Generating new candidates from L_{k-1}//
    forall transactions t∈T' do
        C_t = subset(C_k, t);   // Generating candidate subsets w.r.t. t//
```

```
            forall candidates c ∈ C_t do
                  c.support = c.support + μ_tc
            endfor
      endfor
      L_k = [ c ∈ C_k | c.support ≥ min-support × |T|]
endfor
All frequent itemset = ∪_k L_k
```

Where Apriori-Gen is a procedure to generate the set of all candidate itemsets, which is mainly the join operation and represented as follows [13]:

Candidate Itemsets Generation: without loss of generality, assuming that the items in each itemset are kept sorted in lexicographic order. First, join L_{k-1} with L_{k-1}:

```
insert into C_k
select p.item_1, p.item_2, ..., p.item_{k-1}, q.item_{k-1}
from L_{k-1} p , L_{k-1} q
where p.item_1 = q.item_1, ..., p.item_{k-1} = q.item_{k-2}, p.item_{k-1} < q.item_{k-1};
```

Next, delete all itemsets $c \in C_k$ such that some (k-1)-subset of c is not in L_{k-1}, which implies that the degree of support of c is less than min-support:

```
forall candidates c ∈ C_k do
      forall (k-1)-itemsets d ⊆ c and c ∈ C_k do
            if d ∉ L_{k-1} then
                  delete c from C_k
            endif
      endfor
endfor
```

Finally, we will be able to generate the rules based on the frequent itemsets and their associated degrees of support and degrees of confidence. Specifically, the classical Fast algorithm [3] is extended to take into account the extended notions of Dsupport and Dconfidence due to the introduction of fuzziness in the taxonomies. Concretely, we have sub-algorithm Extended Fast as follows:

Sub-algorithm Extended Fast:

```
forall frequent itemsets l_k, k>1 do
      call gen-rules(l_k, l_k)
endfor

procedure gen-rules(l_k: frequent k-itemset, a_m: frequent m-itemset)
      A={(m-1)-itemsets a_{m-1} | a_{m-1} ⊂a_m}
```

```
        forall a_{m-1} ∈ A do
            conf = Σ_{t∈T}count (μ_{tlk}) / Σ_{t∈T}count (μ_{tam-1})
          if (conf ≥ min-confidence) then
                output the rule a_{m-1}⇒(l_k – a_{m-1}),
                with Dconfidence = conf and Dsupport = l_k.support / |T|
                if (m-1 > 1) then
                      call gen-rules (l_k, a_{m-1})
                endif
          endif
      endfor
  endprocedure
```

where each frequent itemset l_k (or a_m) is associated with $\Sigma_{t\in T}count\ (\mu_{tlk})$ (or $\Sigma_{t\in T}count\ (\mu_{tam})$). In this way, the rules generated all satisfy the pre-specified min-support and min-confidence thresholds.

It should be mentioned, however, that we can discover both $XY \Rightarrow Z$ and $X \Rightarrow Z$, but according to the definition of association rules, we cannot obtain any one from the other. This can be proved as follows:

1) From $X \Rightarrow Z$, i.e., $\|XZ\| / |T| \geq$ min-support, $\|XZ\| / \|X\| \geq$ min-confidence, We have $\|XZ\| \geq \|XYZ\|$ and $\|X\| \geq \|XY\|$. So $\|XZ\| / |T| \geq \|XYZ\| / |T|$, but Dsupport($XY \Rightarrow Z$) = $\|XYZ\| / |T|$ may or may not be $\geq$ min-support. And Dconfidence($XY \Rightarrow Z$) = $\|XYZ\| / \|XY\|$ may or may not be $\geq \|XZ\| / \|X\|$, so we cannot determine whether Dconfidence($XY \Rightarrow Z$) $\geq$ min-confidence. Thus, $XY \Rightarrow Z$ cannot be induced.

2) Likewise, from $XY \Rightarrow Z$, i.e., $\|XYZ\| / |T| \geq$ min-support, $\|XYZ\| / \|XY\| \geq$ min-confidence, we have $\|XZ\| \geq \|XYZ\|$ and $\|X\| \geq \|XY\|$. So $\|XZ\| /|T| \geq \|XYZ\| / |T|$, then Dsupport($X \Rightarrow Z$) = $\|XZ\| / |T| \geq$ min-support. And Dconfidence($X \Rightarrow Z$) = $\|XZ\| / \|X\|$ may or may not be $\geq \|XYZ\| / \|XY\|$, so we cannot determine whether Dconfidence($X \Rightarrow Z$) $\geq$ min-confidence. Thus, $X \Rightarrow Z$ cannot be induced.

To further filter the rules, the extended R-interest measure discussed in section 2, for instance, may be used. Notably the R-interest measure is separated from the process of mining rules, and the method is the same as the classical one proposed by Srikant and Agrawal in [13].

3.2. The Degree of Support and Mining Algorithms: A Fuzzy Implication Viewpoint

As discussed previously, the degree of support, namely Dsupport for rule $X \Rightarrow Y$ is related to $\|X \cup Y\|$, which implies that Dsupport($X \Rightarrow Y$) is the same as Dsupport($Y \Rightarrow X$) in the either crisp or fuzzy case. In the fuzzy case, $\mu_{tX\cup Y}$ is equal to $\min(\mu_{tX}, \mu_{tY})$, i.e., for any t in T,

$$\min(\mu_{tX}, \mu_{tY})$$
$$= \min(\min_{x\in X}(\max_{a\in t}(\mu_{xa})), \min_{y\in Y}(\max_{a\in t}(\mu_{ya})))$$
$$= \min_{z\in Z}(\max_{a\in t}(\mu_{za}))$$
$$= \mu_{tZ}$$

where $X \cup Y = Z$, and μ_{tX} and μ_{tY} are in [0,1]. In the crisp case, $||X \cup Y||$ is counted for those transactions that contain both X and Y. In terms of μ_{tX} and μ_{tY} in {0.1}, both μ_{tX} and μ_{tY} are needed to be 1 in order to be counted in $||X \cup Y||$. Thus, $\mu_{tX \cup Y} = \min(\mu_{tX}, \mu_{tY})$ is also true for the crisp case.

On the other hand, one may try to distinguish between X⇒Y and Y⇒X for Dsupport in some way. A possible attempt is to link X and Y using fuzzy implication operators. In other words, the degree of support for rule X⇒Y is related to the truth value of the fuzzy implication from μ_{tX} to μ_{tY}, i.e., FIO(μ_{tX}, μ_{tY}), where FIO is a fuzzy implication operator [4, 8]. The degree that a transaction t supports rule X⇒Y is therefore denoted as $\mu_{tX\Rightarrow Y}$. Furthermore, $\mu_{tX\Rightarrow Y}$ may be defined as follows:

$$\mu_{tX\Rightarrow Y} = \min(\mu_{tX}, FIO(\mu_{tX}, \mu_{tY}))$$

Here, taking μ_{tX} with FIO(μ_{tX}, μ_{tY}) in the *min* operation is to conform with the semantics that t supporting X⇒Y usually assumes that both X and Y appear in t at the same time (though at different degrees). In accordance with the definition of $\mu_{tX\Rightarrow Y}$, the degree of support for X⇒Y, Dsupport(X⇒Y), and the degree of confidence for X⇒Y , Dconfidence(X⇒Y), can be determined as follows:

$$Dsupport(X\Rightarrow Y) = \sum_{t\in T} count\ (\mu_{tX\Rightarrow Y}) \ / \ |T|$$

$$Dconfidence(X\Rightarrow Y) = \sum_{t\in T} count\ (\mu_{tX\Rightarrow Y}) \ / \ \sum_{t\in T} count\ (\mu_{tX})$$

Moreover, the R-interest measure may be determined in a similar manner. Therefore, as a general setting from the FIO perspective, the classical Fast algorithm is extended in the following way:

Sub-algorithm FIO-Extended Fast:

```
forall frequent itemsets l_k, k>1 do
        call gen-rules(l_k, l_k)
    endfor

    procedure gen-rules(l_k: frequent k-itemset, a_m: frequent m-itemset)
        A={(m-1)-itemsets a_m-1 | a_m-1 ⊂ a_m}
        forall a_m-1 ∈A do
```

```
conf = Σ_{t∈T}count(μ_{tam-1⇒(lk-am-1)}) / Σ_{t∈T}count(μ_{tam-1})
if (conf ≥ min-confidence) then
    output the rule a_{m-1}⇒(l_k − a_{m-1}),
    with Dconfidence = conf and Dsupport = l_k.support / |T|
    if (m-1 > 1) then
        call gen-rules (l_k, a_{m-1})
    endif
endif
endfor
endprocedure
```

where $\mu_{tam-1\Rightarrow(lk-am-1)} = \min(\mu_{tam-1}, FIO(\mu_{tam-1}, \mu_{t(lk-am-1)}))$ is the degree that t supports the rule $a_{m-1}\Rightarrow(l_k - a_{m-1})$. There are a number of fuzzy implication operators (FIOs) that have been studied in literature [4, 8]. The authors are currently conducting a study on some of the FIOs and their properties in the context of FIO-extended Fast algorithm. Although the detailed discussions and technical treatments of the issues go beyond the scope of this paper, two points are worth mentioning. First, the FIO-Extended Fast algorithm embraces the Extended Fast algorithm in the sense that the FIO-Extended Fast algorithm becomes the Extended Fast algorithm when the min operator M (M(a, b) = min(a, b)) is in place of FIO:

$$
\begin{aligned}
&\mu_{tam-1\Rightarrow(lk-am-1)}\\
&= \min(\mu_{tam-1}, FIO(\mu_{tam-1}, \mu_{t(lk-am-1)}))\\
&= \min(\mu_{tam-1}, M(\mu_{tam-1}, \mu_{t(lk-am-1)}))\\
&= \min(\mu_{tam-1}, \min(\mu_{tam-1}, \mu_{t(lk-am-1)}))\\
&= \min(\mu_{tam-1}, \mu_{t(lk-am-1)})\\
&= \mu_{tlk}
\end{aligned}
$$

Second, using FIOs for Dsupport may distinguish Dsupport(X⇒Y) from Dsupport(Y⇒X) to a certain extent, depending on what specific FIOs are chosen. Merely for illustrative purposes, EA fuzzy implication operator (EA(a, b) = max(1-a, min(a, b)) for all a, b in [0, 1], see [4, 8]) is used in the following example to help show the idea.

The example is similar to the example in section 2.4, but with slight changes in the taxonomies and the transactions, which are represented in Figure 3 and Table 5 respectively.

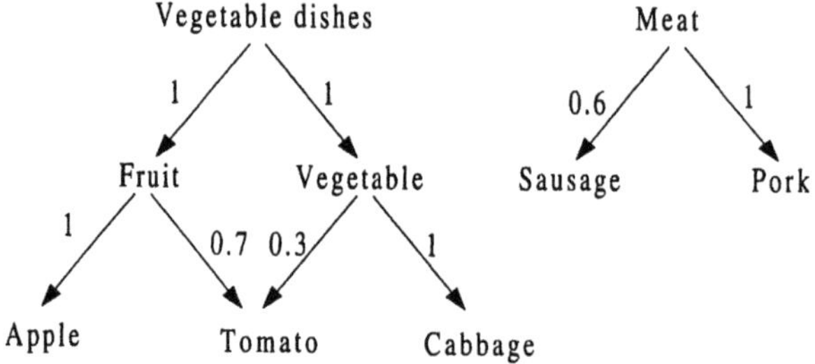

Figure 4 Example of fuzzy taxonomic structures (revised)

Transaction #	Things Bought
#100	Apple
#200	Tomato, Sausage
#300	Cabbage, Sausage
#400	Tomato, Pork
#500	Pork
#600	Cabbage, Pork

Table 5 Transactions in a supermarket database

Again with min-support set to 30%, the frequent itemsets generated are shown in Table 6. Then applying the FIO-Extended Fast algorithm with EA to Table 6 will result in the rules shown in Table 7.

Frequent Itemsets	Σ*count* values
{Cabbage}	2
{Tomato}	2
{Pork}	3
{Fruit}	2.4
{Vegetable}	2.6
{Vegetable dishes}	4.4
{Meat}	4.2
{Vegetable, Meat}	2.2
{Vegetable dishes, Meat}	2.9

Table 6 Σcount values for frequent itemsets

Note that in Table 7 the degrees of support for the rules (Vegetable⇒Meat) and (Meat⇒Vegetable) do differ from each other and are calculated as follows:

X⇒Y	Dsupport(X⇒Y)	Dsupport(X)	Dconf. (X⇒Y)
Vegetable⇒Meat	36.67%	2.6/6	84.62%
Meat⇒Vegetable	38.33%	4.2/6	54.76%
Vegetable dishes⇒Meat	48.33%	4.4/6	65.91%
Meat⇒Vegetable dishes	48.33%	4.2/6	69.05%

Table 7 The rules satisfying min-support

Dsupport(Vegetable⇒Meat)
= (min(max(min(0, 0), 1-0), 0) + min(max(min(0.3, 0.6), 1-0.3), 0.3) + min(max(min(1, 0.6), 1-1), 1) + min(max(min(0.3, 1), 1-0.3), 0.3) + min(max(min(0, 1), 1-0), 0) + min(max(min(1, 1), 1-1), 1)) / 6
= (0 + 0.3 + 0.6 + 0.3 + 0 + 1) /6
= 2.2 / 6
= 36.67%

Dsupport(Meat⇒Vegetable)
= (min(max(min(0, 0), 1-0), 0) + min(max(min(0.6, 0.3), 1-0.6), 0.6) + min(max(min(0.6, 1), 1-0.6), 0.6) + min(max(min(1, 0.3), 1-1), 1) + min(max(min(1, 0), 1-1), 1) + min(max(min(1, 1), 1-1), 1)) / 6
= (0 + 0.4 + 0.6 + 0.3 + 0 + 1) /6
= 2.3 / 6
= 38.33%

The level of difference between Dsupport(X⇒Y) and Dsupport(Y⇒X) (e.g., between Dsupport(Vegetable⇒Meat) and Dsupport(Meat⇒Vegetable)) relies on the fuzziness in the taxonomies and on the specific FIOs used in the algorithm. Moreover, the rules in Table 7 may be further filtered according to pre-specified min-confidence and R-interest measures.

4. A Preliminary Experiment

This section will reveal some results from a preliminary experiment in that both the classical algorithm and the extended algorithm presented in this article were run on a set of randomly generated synthetic data, which had the parameters as listed in Table 8. By a preliminary experiment we mean the experiment that was meant to illustrate some aspects of the algorithms to certain extents, and carried out with a limited volume of data and in a less powerful computing environment. Notably, more intensive explorations on such aspects of our work as the analysis of computational complexity, the technical treatment of the algorithms, and the experiments with a more vast volume of both synthetic and real data in a more powerful computing environment are being undertaken and will be reported in a separate paper.

Parameters	Description
|T|	Number of transactions
|LN|	Number of leaf items in the taxonomies
|IN|	Number of interior items in the taxonomies
Min-support	Value of min-support
Min-confidence	Value of min-confidence

Table 8 Parameters

For the purpose of comparison, both the classical algorithm (denoted as Classical) and the extended algorithms (denoted as Extended for the algorithm with sub-algorithm Extended Fast, and denoted as EA-Extended for the algorithm with sub-algorithm FIO-Extended Fast with EA) were implemented. These three algorithms, namely Classical, Extend and EA-Extended, were run with up to 1,000,000 transactions. The experiment was carried out using a personal computer with Intel MMX166/16M RAM with Microsoft Foxpro 2.5B.

In fact, compared with the classical algorithm in [3, 13], we can find there are two differences between classical algorithm and extended algorithms. The first is about Sub-algorithm Degree. In the Extended algorithms, we should compute the degree between each leaf node and its ancestor, while we do not need to do so in the classical algorithm. The second is that we replace *count* operation with *Σcount* operation. However, such differences would not affect the efficiency of the algorithm much. As a matter of fact, in terms of algorithms' structures and loops, etc., the extended algorithms are of the same level of computational complexity as that of the classical algorithm. The experimental results also conform to it.

Number of transactions. In the experiment, the number of transactions varied from 100 to 1,000,000. In order to test the influence of |T|, min-support is set to 0. The |T|-Time is shown in Figure 4.

Figure 5 shows that the three curves are almost overlapping, especially when |T| gets larger. This also reveals, to a certain degree, that given a fuzzy taxonomic structure, the Sub-algorithm Degree affects the efficiency little. Moreover, the algorithms perform equally well and are polynomial in efficiency with respect to |T|.

Number of leaf items in the taxonomies. Because all the algorithms perform heavily on the manipulation of subsets and operations of join, they are expected to be exponential with respect to the number of items, which is reflected in Figure 6. In addition, both the classical algorithm and the extended algorithms show the same level of performance. Note that the performance may improve as min-support and min-confidence increase. Because with the increase of min-support, more and more k-candidate itemsets are filtered. Thus, (k+1)-candidate itemsets generated on the k-frequent itemsets are less and less.

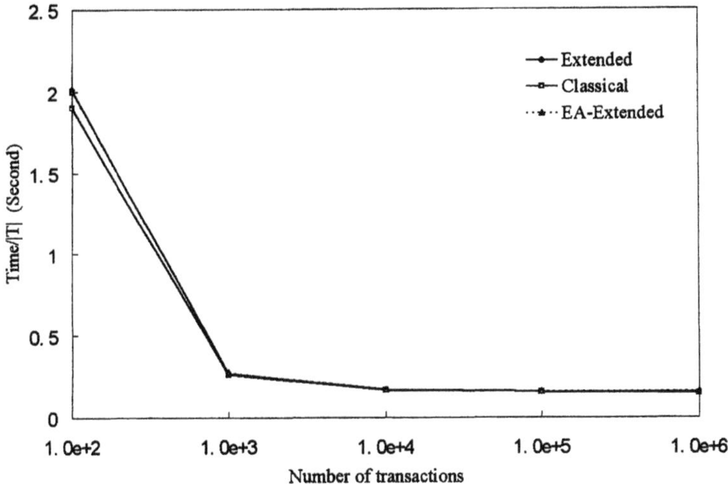

Figure 5 Number of transactions (with |LN| = 10 and |IN| = 4)

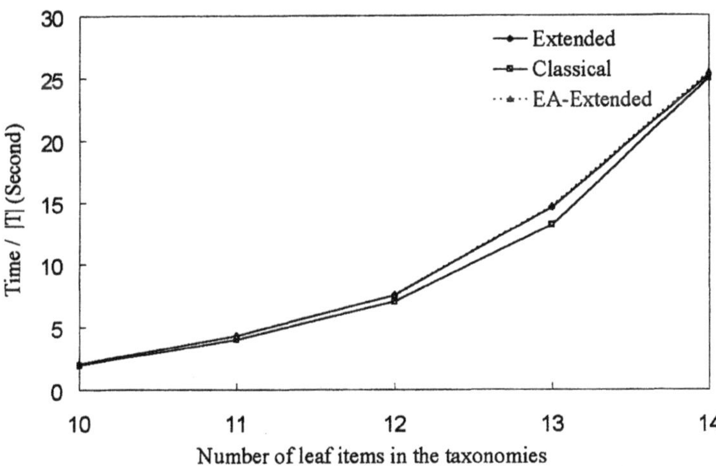

Figure 6 Number of leaf items in the taxonomies (min-support =0 and min-confidence= 0).

Number of the interior items in the taxonomies. The extended algorithms are expected to rely more on the number of the interior items than the classical algorithm since the extended algorithms require more computation and join operations for the partial degrees of the nodes due to fuzzy taxonomies. This is reflected in Figure 7 by the distance between the Classical curve and the

Extended/EA-Extended curves. It appears that the distance is getting larger as the number of interior items increases, which conforms to the intuition.

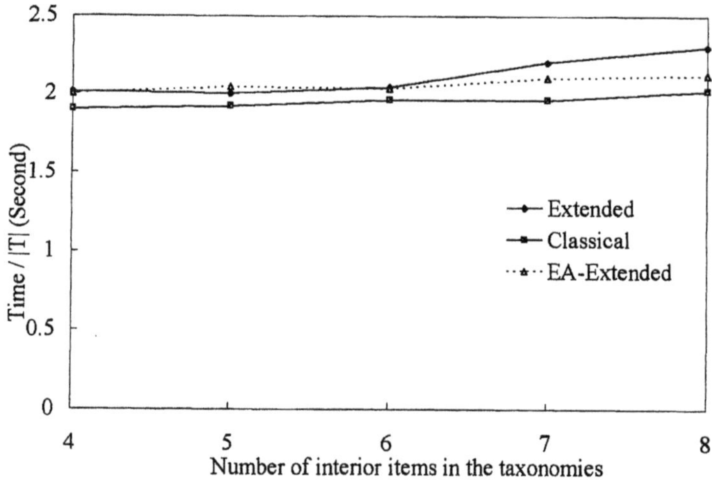

Figure 7 Number of interior items in the taxonomies

Min-support. In examining other parameters (e.g., figures 5, 6 and 7), min-support was set to 0. As mentioned previously, higher min-support may help the algorithms perform better. Figure 8 reflects this fact.

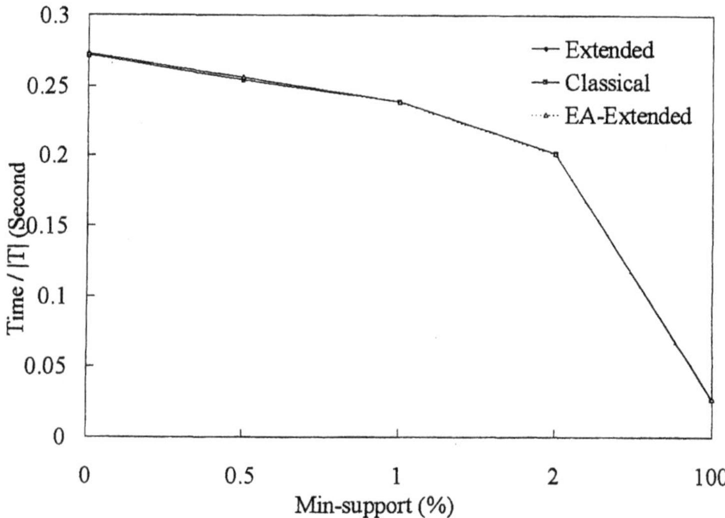

Figure 8 Min-support

5. Conclusions and Further Studies

Aimed at dealing with the taxonomic inexactness when mining generalized association rules, this paper has introduced the fuzziness in the underlying taxonomic structures and extended the classical algorithm in a way that a transaction may partially support a particular item. This has then led to re-examining the computation for the degree of support and the degree of confidence, as well as for the R-interest measure. Furthermore, the classical algorithm of Srikant and Agrawal's (including the Apriori algorithm and Fast algorithm) has been extended to incorporate the extended notions of Dsupport, Dconfidence and R-interest. In so doing, a number of sub-algorithms have been developed, namely, sub-algorithm Degree, sub-algorithm Extended Transaction Set T', sub-algorithm Extended Apriori, sub-algorithm Extended Fast, and sub-algorithm FIO-Extended Fast. The FIO-Extended Fast sub-algorithm is a general setting in an attempt at distinguishing the degree of support for $X \Rightarrow Y$ from that for $Y \Rightarrow X$. Some examples have been provided to help illustrate the ideas. Moreover, the results of a preliminary experiment have shown that the extended algorithms performed almost equally well in $|T|$ compared with the classical algorithm, and that the extended algorithms relied more on the number of interior items than the classical algorithm due to the introduction of partial belongings in the taxonomic structures.

The fuzzy extensions presented in this article enable us to discover not only crisp generalized association rules but also fuzzy generalized association rules within the framework of fuzzy taxonomic structures. Our ongoing and future studies include the detailed treatment and analysis of the extended algorithm (sub-algorithms) and other related algorithms, more tests on the computational complexity with a large amount of synthetic and real data, explorations of FIOs in the context of FIO-Extended Fast sub-algorithm, and further extensions that allow discovering more general forms of fuzzy association rules which are meaningful and important to the decision makers.

References

1 Rakesh Agrawal, Tomasz Imiclinski, Arun Swami, *Mining Association Rules between Sets of Items in Large Databases*, Proceedings of the 1993 ACM SIGMOD Conference Washington DC, USA, May 1993.

2 Rakesh Agrawal, Heikki Mannila, Ramakrishnan Srikant, Hannu Toivonen, A. Inkeri Verkamo, *Fast Discovery of Association Rules* in Advances in Knowledge Discovery and Data Mining, AAAI Press/The MIT Press, 1996.

3 Rakesh Agrawal, Ramakrishnan Srikant, *Fast Algorithms for Mining Association Rules*, Proceedings of VLDB Conference, Santiago, Chile, Sept. 1994. Expanded version available as IBM Research Report RJ9839, June 1994.

4 Guoqing Chen, Fuzzy Logic in Data Modeling: semantics, constraints and database design, Kluwer Academic Publishers, Boston, 1998.

5 Takeshi Fukuda, Yasuhiko Morimoto, Shinichi Morishita, Takeshi Tokuyama, *Data Mining Using Two Dimensional Optimized Association Rules: Scheme, Algorithm and Visualization*, SIGMOD'96 6/96 Montreal Canada, 1996.

6 J. Han, Y. Fu, *Discovery of Multiple-level Association Rules from Large Databases*, Proceedings of the 21st International Confernce on Very Large Databases, Zurich, Switzerland, September 1995.

7 Maurice Houtsma, Arun Swami, *Set-oriented Data Mining in Relational Databases*, Data & Knowledge Engineering, 17 (1995) 245-262.

8 Etienne E. Kerre, Introduction to Basic Principles of Fuzzy Set Theory and Some of Its Applications. 2nd edition. Gent, Belgium: Communication & Cognition, 1993.

9 Mika Klemettinen, Heikki Mannila, Pirjo Ronkainen, Hannu Toivonen, A. Inkeri Verkamo, *Finding Interesting Rules from Large Sets of Discovered Association Rules*, Proceedings of Third International Conference on Information and Knowledge Management, Nov. 29-Dec. 2, 1994.

10 Heikki Mannila, Hannu Toivonen, A. Inkeri Verkamo, *Efficient Algorithms for Discovering Association Rules*, AAAI Workshop on Knowledge Discovery in Databases, pp. 181-192, Seattle, Washington, July 1994.

11 Edward Minieka, Optimization Algorithms for network and Graph, Marcel Dekker, Inc., New York and Basel, 1978.

12 Savasere, E. Omiecinski, S. Navathe. *An Efficient Algorithm for Mining Association Rules in Large Databases*, Proceedings of the VLDB Conference, Zurich, Switzerland, September 1995.

13 Ramakrishnan Srikant, Rakesh Agrawal, *Mining Generalized Association Rules*, Proceedings of the 21st VLDB Conference Zurich, Swizerland, 1995.

14 Ramakrishnan Srikant, Rakesh Agrawal, *Mining Quantitative Association Rules in Large Relational Tables*, SIGMOD'96 6/96 Montreal, Canada, 1996.

15 Weimin Yan, Weimin Wu, Data Structure, Tsinghua University Press, 1992.

On Combining Intelligent Querying and Data Mining Using Fuzzy Logic Concepts

Janusz Kacprzyk and Slawomir Zadrozny

Systems Research Institute, Polish Academy of Sciences
01-447 Warsaw, ul. Newelska 6, Poland

Abstract
In this contribution we discuss how a fuzzy querying interface can support the generation of linguistic database summaries - a special technique of data mining. Links between our approach to linguistic summaries and the well-known technique of association rules is shown. The implementation of linguistic summaries generation using the authors' FQUERY for Access package is presented.

Keywords: fuzzy querying, linguistic terms, data mining, database summaries, association rules

1 Introduction

Fuzzy querying of traditional databases is widely perceived as the most promising area of fuzzy logic application in database theory. A lot of theoretical and practical results have been obtained [1-13]. Fuzzy querying interface makes it possible for a user to employ his or her own dictionary while composing a query. Thus, there are much higher chances that the query keeps its original meaning as conceived by the user. For example, a customer of a real-estate agency looking for a house would rather express his or her criteria using imprecise descriptions as *cheap*, *large* garden, etc. Also, to specify which combination of the criteria fulfillment would be satisfactory, he or she would often use expressions like, *most* of them or *almost all* (should be fulfilled). Such vague terms may be relatively easily processed using fuzzy logic. This has motivated the development of a whole family of fuzzy querying interfaces, notably the FQUERY for Access package proposed and implemented by the authors [4-10].

The same linguistic paradigm may be applied in the framework of data mining. Firstly, linguistic, imprecise terms used to represent rules, patterns etc. may much more easily account for noisy, incomplete etc. data. Secondly, using a linguistic representation for pieces of discovered knowledge, we can immediately present it to the end-user, without any additional representation transformations. For example, a database summary like "*most* our customers are *reliable*" may be more useful than, say "65% of our customers have paid at least 70% of their duties in less than 10 days".

From this perspective, *fuzzy* (*linguistic*) queries are directly applicable for the data mining purposes [14, 15]. Both the approaches may be regarded as dealing with the same building blocks. Moreover, the data mining efforts may be guided by the recorded history of fuzzy queries posed against a given database.

The organization of the paper is as follows. First, we present the basics of the calculus of linguistically quantified proposition. Such propositions are crucial for both fuzzy querying and data mining. Secondly, we briefly present the main points of our fuzzy querying concept. Then, we discuss how fuzzy querying interface may support the generation of database summaries - a special technique of data mining. The next section shows links between our approach to linguistic summaries and the well-known technique of association rules [16]. Finally, we briefly present the implementation of generation of linguistic summaries using our FQUERY for Access package.

We use a standard relational database management systems terminology, i.e. a database contains *tables*, which are defined by a set of *attributes* (*columns*), and contain *rows*.

2 A Calculus of Linguistically Quantified Propositions

A calculus of linguistically quantified propositions has been introduced by Zadeh [17, 18]. Basically, it makes it possible to assess the truth-value of a proposition of the following type:

"Most elements of set X posses property S" (1)

that may be formally expressed as follows:

$$\underset{x \in X}{Q} \; S(x) \tag{2}$$

where Q denotes a fuzzy linguistic quantifier (e.g., "most"), $X = \{x_1, \dots x_m\}$ is a universe of discourse, and $S(\cdot)$ is a property which is assumed fuzzy and its interpretation may be informally equated with a fuzzy set, i.e.:

$$\text{truth}(S(x_i)) = \mu_S(x_i)$$

In the framework of this calculus, propositions of the following type are also considered:

"Most elements of set X possessing property F posses also property S" (3) that may be formally expressed as follows:

$$\underset{x \in X}{QF}\, S(x) \qquad (4)$$

Property F is here assumed fuzzy too, and may be equated with a fuzzy set F, i.e.:

$$\text{truth}(F(x)) = \mu_F(x)$$

A linguistic quantifier is represented as a fuzzy set Q in [0,1]. Informally, each number from this interval represents what fraction of the elements of the universe of discourse satisfies the proposition. Then, the membership function of a quantifier is defined so as to represent its semantics. For example, in case of the "most" quantifier, the higher the fraction the better, and this may be expressed by the membership function defined by (5) and sketched in Figure 1.

$$\mu_Q(y) = \begin{cases} 1 & \text{for } y \geq 0.8 \\ 2y - 0.6 & \text{for } 0.3 < y < 0.8 \\ 0 & \text{for } y \leq 0.3 \end{cases} \qquad (5)$$

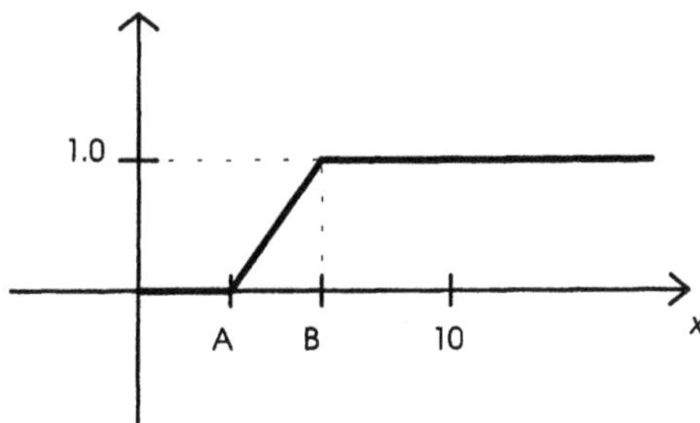

Figure 1. The membership function of the fuzzy linguistic quantifier "*most*"

In the context considered here, we are interested in monotonous quantifiers only exemplified by "most" and characterized by a monotonous membership function, μ_Q, assumed to be piece-wise linear, as in Figure 1.

The following formula is employed to determine the truth of the proposition (2):

$$\text{truth}(QS(X)) = \mu_Q(\sum \text{Count}(S) / \sum \text{Count}(X)) = \mu_Q(\sum_{i=1}^{m} \mu_S(x_i) / m) \qquad (6)$$

where m = card(X) and $\sum \text{Count}(A)$ refers to the cardinality of a fuzzy set A, given by:

$$\sum \text{Count}(A) = \sum_{x_i \in X} \mu_A(x_i)$$

In case of the proposition (4) the following formula is employed:

$$\text{truth}(QF\ S(X)) = \mu_Q(\sum \text{Count}(S \cap F) / \sum \text{Count}(F)) =$$

$$= \mu_Q(\sum_{i=1}^{m}(\mu_S(x_i) \wedge \mu_F(x_i)) / \sum_{i=1}^{m} \mu_F(x_i)) \tag{7}$$

where, $\sum_{i=1}^{m} \mu_F(x_i) \neq 0$, and $\wedge$ is a *t*-norm (see, e.g., [19]).

Example

Let $X = \{x_1, \ldots x_{10}\}$ be a set of the employees of a company. Let *S*(*x*) and *F*(*x*) represent the properties "*x*'s salary is *low*" and "*x* is *young*", respectively, and the corresponding fuzzy sets be:

	x_1	x_2	x_3	x_4	x_5	x_6	x_7	x_8	x_9	x_{10}
μ_S	0.8	1	1	0.2	0.3	0.2	0.5	0	0	0
μ_F	1	1	1	0.8	0.6	0.5	0.3	0	0	0

Then, assuming *most* represented with (5), we can calculate the truth-value of the following two propositions: "*Most* employees's salaries are *low*" and "*Most young* employees' salaries are *low*". Using (6) and (7) we obtain:

truth("*Most* employees' salaries are *low*")=0.4
truth("*Most young* employees' salaries are *low*")=3.8/5.2=0.73

3 The Concept of Fuzzy Querying

The idea of fuzzy querying that we have adopted consists in a direct use of linguistic terms in queries. It is widely argued that for a human being natural language is a natural medium to form and express thoughts. Obviously, it also applies to a query that may be treated as a set of search criteria conceived by a user. Thus, the possibility to use elements of a natural language should make the querying of a database to be much more a comfortable experience for an average user. This is especially promising in case of a frequent use of a database by the same user. In such a situation the querying interface may learn more effectively the dictionary of given user. It often happens that the same words mean something completely different for distinct humans. In case of one user and a rather limited universe of discourse, as defined by the subject of the database, one can expect much more stable semantics of the user's dictionary.

Often the meaning of linguistic terms is vague. This implies that we have to adopt some special formal means to deal with them. Moreover, it would be unreasonable to require the answer for a vague query to be completely precise, adhering to the classical yes-no logic. Thus, instead of listing the crisp set of matching database rows, a fuzzy query against precise data yields a *fuzzy* set of such rows. Namely, each listed row is accompanied by its *matching degree* against the query.

In our approach, we employ fuzzy logic to represent the meaning of the linguistic terms used in queries. The following types of linguistic terms are assumed:

- *numerical fuzzy values*, exemplified by *low* in "profitability is *low*",
- *scalar fuzzy values*, exemplified by *Central Europe* in "country is in *Central Europe*",
- *fuzzy relations*, exemplified by *much greater than* in "income is *much greater than* spending", and
- *linguistic quantifiers*, exemplified by *most* in "*most* conditions have to be met".

These elements are building blocks of fuzzy queries in our approach. All these linguistic terms, similarly to linguistic quantifiers, are represented as fuzzy sets. Scalar fuzzy values are for obvious reasons dedicated for specific attributes. For example, an imprecise term "*Central Europe*" is meaningful only in the context of attributes related to the country. Thus, "*Central Europe*" has to be defined as a fuzzy set in the (finite) space of all values of given attribute by, e.g., the following membership function:

1.0/CzechRepublic + 1.0/Hungary + 1.0/Poland +
+ 1.0/Slovakia + 0.8/Austria + 0.6/Ukraine + ...

On the other hand, in our approach *numerical fuzzy values* may be applicable for any numerical attribute, provided it is declared what is the range of values this attribute takes. These fuzzy values are defined as fuzzy sets on a universal interval [-10, +10] and are converted in an appropriate way when used along with given numerical attribute. This does not prevent the definitions of numerical fuzzy values specific for particular attributes.

Usually, a query consists of a number of simple conditions combined using the classical logical connectives AND and OR. In our approach, *linguistic quantifiers* provide for a more flexible aggregation scheme of simple conditions in queries. For example, instead of requiring that all simple conditions are met, using an appropriate linguistic quantifier one may indicate that *most* of them are to be met.

Before a linguistic term may be used in a query, it has to be defined and stored internally. The definition consists of a label of the term and a membership function of the associated fuzzy set. This feature, i.e. maintenance of *dictionaries* of linguistic terms defined by users as well as predefined in the system, supports both fuzzy querying and data mining, to be discussed next.

This concludes the brief presentation of our concept of fuzzy querying. It has been implemented in our software, the FQUERY for Access package. It is an add-in to Microsoft Access providing the user of this popular desktop database management system with fuzzy querying capabilities. For more information on the software implementation and underlying querying concept consult a series of publications by the present authors [4-10].

4 Data Mining through Linguistic Database Summaries

Data mining, also known as knowledge discovery in databases, deals with finding interesting patterns, dependencies, regularities etc. in bodies of data stored in databases. Yager [14] proposed to use linguistically quantified propositions to *summarize* the content of a database, i.e. to grasp characteristic features of analyzed data. Here we follow and extend this idea in the context of a fuzzy querying concept briefly presented in the previous section.

We start by restating the primary concepts of the calculus of linguistically quantified propositions in terms relevant for database summaries. Thus, (1) becomes:

"*Most rows match query S*" (8)

Similarly, (3) becomes:

"*Most rows meeting conditions F match query S*" (9)

The query is meant here as a collection of conditions on the values of attributes combined using some logical connectives (including linguistic quantifiers). These conditions may refer to various linguistic terms mentioned in the previous section. Effectively, a query *S* defines a fuzzy subset (fuzzy property) on the set of the rows, where the membership of them is determined by their matching degree with the query. Thus, (8) states that most of the rows posses a certain property defined by a query - possibly an interesting piece of knowledge about data. The more complex is the query involved, the less trivial is the summary obtained. In large data sets it may relatively seldom happen that most of the rows satisfy a given non-trivial query. Then, a proposition of type (9) may be helpful. Namely, it states that a subset of rows satisfies conditions specified in a query. In the database terminology, *F* corresponds to a *filter* and (9) asserts that *most* rows passing through *F* match query *S*. Moreover, since the filter may be fuzzy, a row may pass through it to a degree from [0,1]. As this is more general than (8), we will assume (9) as a basis for our further considerations.

In Yager's approach [14] the truth-value of a given linguistically quantified proposition of type (8) or (9) is the primary measure of quality of the obtained summary. Practically, such a basic quality indicator has to be supplemented with some measure of "interestingness" ("non-triviality", "unexpectedness", ...). In what follows, we will discuss different types of linguistic data summaries based on the general scheme given by (9).

Thus, the generation of a linguistic summary consists in finding propositions of type (9) that have a high truth-value. Basically, a proposition sought consists of three elements: a fuzzy filter F (optional), a query S, and a linguistic quantifier Q. There are two limit cases, where we:

- do not assume anything about the form of any of these elements,
- assume concrete forms of a fuzzy filter and query, and look only for a linguistic quantifier Q.

Obviously, in the first case the process of data summarization will be extremely time-consuming but may produce interesting results, not predictable by the user in any other way. In the second case the user somehow has to guess a good candidate formula for the summarization but the evaluation is fairly simple as it requires more or less only the same resources as the answering of a (fuzzy) query. Thus, the second case refers to the summarization known as *ad hoc queries*, extended with an automatic determination of a linguistic quantifier.

In-between these two extreme cases there are different types of summaries, with various assumptions on what is given and what is sought. In case of a linguistic quantifier the situation is simple: it may be given or sought. In case of a fuzzy filter F and a fuzzy query S, more possibilities exist. Basically, both F and S consist of simple conditions, each stating what *value* an *attribute* should take on, and connected using logical connectives. Here we assume that the query refers to the attributes from one, fixed table. Obviously, such a table may be prepared by joining several source tables before the summary is started. An automatic selection of tables to be considered by summaries seems to be very complicated. It would require some additional meta-data on data stored in a database and/or some user's intervention during the generation of a summary. Some useful meta-data comes from integrity constraints - typically identified during a relational database design and encoded within some system tables. Still, we assume here only one table of interest.

We will use the following notation to describe what is given or what is sought with respect to the fuzzy filter F and query S (below A stands for either F or S):

- A - all is given (or sought), i.e., attributes, values and the structure,
- A^{fc} - attributes and structure are given, but values are left out,
- A^{v} - denotes sought left out values referred to in the above notation, and
- A^{f} - only a set of attributes is given, the other elements are sought.

Using the above notation we may propose the following classification of the summaries:

Type	Given	Sought	Remarks
1	S	Q - sought	Simple summaries through ad-hoc queries
2	$S\ F$	Q - sought	Conditional summaries through ad-hoc queries
3	$Q\ S^{fc}$	S^{v}	Simple value oriented summaries
4	$Q\ S^{fc}\ F$	S^{v}	Conditional value oriented summaries
5	nothing	$S\ F\ Q$	General fuzzy rules

Thus, we distinguish 5 main types of data summarization. Type 1 may be easily produced by a simple extension of fuzzy querying as proposed and implemented in our FQUERY for Access package, see Section 6 for more details. Basically, the user has to construct a query - a candidate summary. Then, it has to be determined what is the fraction of the rows matching this query and what linguistic quantifier best denotes this fraction. The primary target of this type of summarization is certainly to propose such a query that a large proportion, e.g., *most*, of the rows satisfies it. On the other hand, it may be interesting to learn that only *few* rows satisfy some meaningful query. A Type 2 summary is a straight extension of Type 1 summaries by adding a fuzzy filter. As soon as a fuzzy querying engine deals with fuzzy filters, the computational complexity of this type of summaries is the same as for Type 1. For more on these types of summaries, see for instance [8], and [20] for a non-fuzzy approach.

The summaries of Type 3 require much more effort. A primary goal of this type of summary is to determine typical (exceptional) values of an attribute. In such a special case, query S consists of only one simple condition built of the attribute whose typical (exceptional) value is sought, the '=' relational operator and a placeholder for the value sought. For example, using the following summary in a context of personal data:

Q = "*most*" and S = "age=?" ("?" denotes placeholder mentioned above)

we look for a typical value of the age of the employees.
Then, we try to find such a (possibly fuzzy) value that the query matches to a high degree Q of the rows. Depending on the category of the Q used as, e.g., *most* versus *few*, typical or exceptional values are sought, respectively. Some more considerations are required as in some cases all values may turn out to be exceptional and none to be typical. This type of summaries may be used with more complicated, regular queries but it may quickly become computationally infeasible (due to the combinatorial explosion) and the interpretation of results becomes vague. A Type 4 summary may produce typical (exceptional) values for some, possibly fuzzy, subset of rows. From the computational point of view, the same remarks apply as for Type 1 versus Type 2 summaries.

A Type 5 summary represents the most general form considered here. In its full version this type of summaries is to produce fuzzy rules describing dependencies between specific values of particular attributes. Here the use of the filter is essential, in contrast to the previous types where it was optional. The very meaning of a fuzzy rule obtained is that if a row meets a filter's condition, then it meets also the query's conditions - this corresponds to a classical IF-THEN rule. For a general form of such a rule it is difficult to devise an effective and efficient generation algorithm. Full search may be acceptable only in case of restrictively limited sets of rule building blocks, i.e. attributes and their possible values. Here, some genetic algorithm based approaches are often employed [21]. In order to alleviate the computational complexity, additional assumptions may also be made. For example, some sets of relevant (interesting, promising, etc.) attributes for the query (S^f) and the filter (F^f) may be selected in advance. Some constraints may also be put on the structure of the query S and filter F (in terms of the number of logical connectives allowed). Another, important special case of Type 5 summaries refers to the situation where the query (S) is fixed and only the filter (F) and quantifier (Q) are sought, i.e. we look for causes of given data features. For example, we may set in a query that profitability of a venture is *high* and look for the characterization of ventures (rows) securing such a high profitability.

The summaries of Type 1 and 3 have been implemented as an extension to our FQUERY for Access - see Section 6 for more details.

In the next section we discuss in a more detailed way one more special case of Type 5 summary, for which computationally efficient algorithms are known in the literature.

5 Association Rules and Fuzzy Summaries

The concept of data mining presented in the previous section features an attractive uniform theoretical background referring to the calculus of linguistically quantified propositions. This paves the way for a consistent use of linguistic terms in database summaries. Unfortunately, in the original Yager's version as well as in the previous section there are rather limited clues as to the devising of efficient algorithms generating summaries of large data sets. Simplest types of summaries are directly computationally tractable (see Section 6). Another general approach calls for the use of genetic algorithms [21]. Thus, in this section looking for efficient algorithms we investigate the similarity of our Type 5 summaries to what is known in the literature as *association rules* [16].

Originally, the association rules were defined for binary valued attributes in the following form:

$$A_1 \wedge A_2 \wedge \ldots \wedge A_n \rightarrow A_{n+1} \tag{10}$$

Thus, such an association rule states, that if in a database row all the attributes from the set $\{A_1, A_2, \ldots, A_n\}$ take on value 1, then also the attribute A_{n+1} is expected to take on value 1 (remember that all attributes are assumed to take on their values from the set $\{0,1\}$).

A row in a database (table) is said to *support* a set of attributes $\{A_i\}_{i \in I}$ if all attributes from the set take on in this row value 1. There are two measures of the quality of an association rule used:

- the *support* of a rule (10) is the fraction of the number of rows supporting the set of attributes $\{A_i\}$, $i\in\{1,..,n+1\}$ in a database (table), and
- the *confidence* of a rule in a database (table) is the fraction of the number of rows supporting the set of attributes $\{A_i\}$, $i\in\{1,..,n+1\}$ among all rows supporting the set of attributes $\{A_i\}$, $i\in\{1,..,n\}$.

Thus, while the support determines a statistical significance of a rule, the confidence measures its strength in the database. Usually, we are interested in rules having values of the support measure above some minimal threshold and a high value of the confidence measure. This form of association rules was motivated by their early applications for a so-called customer's basket analysis.

A number of efficient algorithms for finding all association rules possessing a required support measure were devised, see, e.g. [16, 22]. Due to that fact, association rules have become a very popular tool for data mining. The original concept of the association rule essentially evolved over time but still the same algorithms are applicable. The extensions to the initial form of the association rule include:

- the right-hand side, like the left-hand side, may contain a conjunction of the attributes instead of just one attribute,
- many-valued scalar values and their hierarchies may be used in rules [23],
- numerical, real-valued attributes may be used in rules [24]
- some constraints may be imposed on combinations of attributes used in rules [25]

Let us check how the concept of the association rule fits into our general format of summaries of Type 5 mentioned in the previous section.

The left-hand and right-hand sides correspond to the filter F and the query S, respectively. In our fuzzy querying framework of Section 3, a query may be composed of a set of simple conditions combined using the classical AND and OR logical connectives as well as linguistic quantifiers. Thus, the structure of the filter and the query available in case of association rules is rather limited but this simplicity secures the existence of efficient algorithms for rules generation.

The truth-value of the summary, defined by (7), corresponds to the confidence measure of a rule. Both definitions become identical when we assume the linguistic quantifier in (7) to be a so-called uniform quantifier characterized by the following membership function: $\mu(y)=y$, for each $y\in[0,1]$.

Thus, in fact an association rule may be treated as a special case of our Type 5 summary. The structure of the filter and the query is limited - only the conjunction of simple conditions is allowed but the number of these conditions and the attributes involved are sought.

A fuzzy querying interface may offer practical ways to circumvent some problems the "classical" [16, 24, 25] search for the association rules faces. For example, when dealing with a numerical attribute, one has to partition its domain into a number of intervals. Then, treating each interval as an additional binary attribute, one can employ known algorithms for the generation of association rules. The question remains how to partition the domain. Srikant and Agrawal [24] advocate the partition into equal intervals but admit that in some cases this may fail. Miller and Yang [26] claim that cluster analysis should be used to determine a proper partition. In fact they go even further by proposing an extension of the association rule concept in the following way. They suggest that it may be advantageous to group attributes provided the distance in corresponding multi-dimensional spaces may be defined in a meaningful way. Then, in a simple condition of a rule - instead of requiring that a numerical attribute lies in given interval - we may require that the values of a group of attributes lie in a cluster.

The use of numerical fuzzy values itself alleviates the problem of a proper partition of an attributes domain. The domain may be covered by a number of overlapping fuzzy values making the partitioning more flexible. Lee and Lee-Kwang [27] discuss the use of classical algorithms to produce association rules referring to fuzzy values. A fuzzy querying interface makes it even simpler as we have a dictionary of fuzzy values at hand, ready to be used in the rules. Moreover, these fuzzy values are familiar for the user as they were defined and tested by him or her in some fuzzy queries. Thus, the meaning of rules produced using such fuzzy values should be much more clear to the user. Context independence of numerical fuzzy values, mentioned in Section 3, is here of special importance as all these values may be used along with all attributes.

The benefit of data mining in the framework of a fuzzy querying interface comes also from that a part of software modules may be useful for both tasks. The simplest summaries may be produced as a side effect of running a query - as it is briefly described in the next section.

6 Data Mining via FQUERY for Access

The concept of fuzzy querying presented in Section 3 has been implemented by the present authors in the FQUERY for Access package. Thus, it extends the querying capabilities of Microsoft Access by making it possible to handle linguistic terms in the queries. This alone makes it an interesting tool for data mining. The simplest method of data mining through ad-hoc created queries becomes much more powerful by using linguistic terms. Nevertheless, the implementation of various types of summaries mentioned in Section 3 seems to be worthwhile. Fortunately, by the very design of FQUERY for Access, such an implementation is relatively straightforward.

We rely on dictionaries of linguistic terms maintained and extended by the users during subsequent sessions. The main feature supporting an easy generation of summaries is the adopted concept of context independent definitions of particular linguistic terms. Hence, looking for a summary we may employ any term in the context of any attribute. Thus, we have got a summary building blocks at hand and what is needed is an efficient procedure for their composition, compatible with the rest of the fuzzy querying system.

In case of Type 1 summaries only the list of defined linguistic quantifiers is employed. The user provides a query S and what is sought is a linguistic quantifier describing in a best way the fraction of rows satisfying this query. Hence, we are looking for a fuzzy set in the space of linguistic quantifiers, such that:

$$\mu_\Sigma(Q) = \text{truth}(Q\ S(x)) \ = \ \mu_Q(\sum_{i=1}^{m} \mu_S(x_i)/m) \tag{11}$$

where $X = \{x_1,\ldots,x_m\}$ is the set of rows.

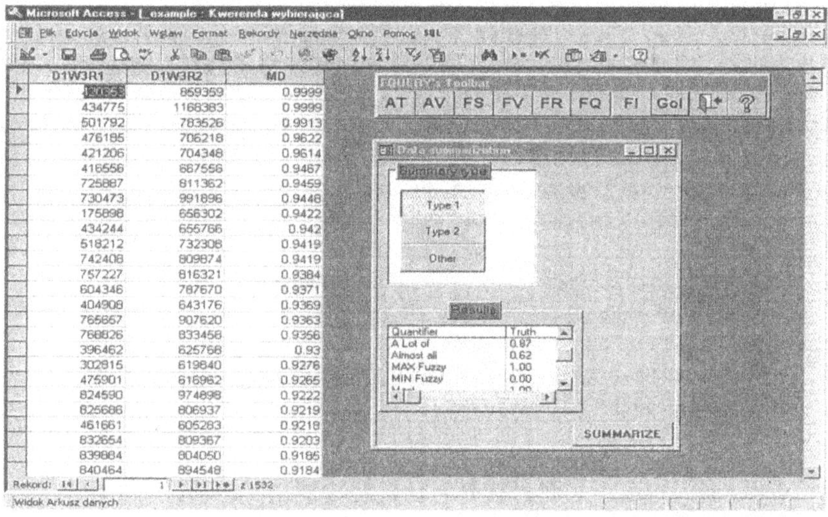

Exhibit 1. Results of a Type 1 summary

FQUERY for Access processes the query in the usual way additionally summing up the matching degrees for all rows. Thus, the sum in (11) is easily calculated. Then the results of the query, i.e. rows accompanied by a matching degree, are displayed in the usual form. In another window, the sought fuzzy set of linguistic quantifiers is shown, as it is shown in Exhibit 1. Currently, FQUERY for Access does not support fuzzy filters. As soon as this capability is added, also summaries of Type 2 will be available. Simply, when evaluating the filter for particular rows, another sum used in (7) will be calculated and the final results will be presented as in case of Type 1 summaries.

Type 3 summaries require more effort and a redesign of the query results display. Now, we are given a quantifier (currently the "most" quantifier is assumed) and the whole query but without some values. Thus, first of all, we have to extend the syntax of the query language introducing a *placeholder* for a fuzzy value. That is, the user may leave out some values in the query's conditions and ask the system to find a best fit for them. During the query processing this left out values are treated in the same way as regular numerical fuzzy values. However, the matching degree is calculated not just for one fuzzy value but for all fuzzy values defined in the dictionary. The matching degrees of the whole query against the subsequent rows, calculated for different combination of fuzzy values are summed up. Finally, it is computed for particular combinations of fuzzy values how well the query is satisfied when a given combination is put into the query. Thus, we again receive as a result a fuzzy set but this time defined in the space of fuzzy values' vectors, as it is shown in Exhibit 2.

Obviously, such computations are extremely time-consuming and are practically feasible only for one placeholder in a query. On the other hand, the case of one placeholder, corresponding to the search for typical or exceptional values, seems to be most useful form of a Type 3 summary.

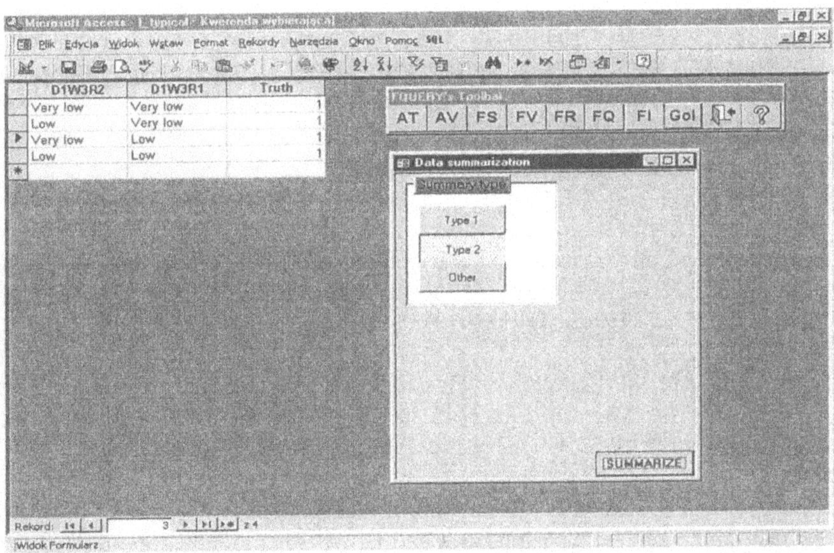

Exhibit 2. Results of a Type 3 summary

7 Concluding Remarks

Combining fuzzy querying with a data mining techniques may produce interesting tools. The use of fuzzy (linguistic) terms makes the data mining results more human-consistent and comprehensible. Moreover, the building blocks of fuzzy queries - fuzzy values and linguistic quantifiers - are readily available for the inclusion into the summaries or association rules. The further research on adaptation and extension of efficient association rules generation algorithms for the case of linguistic setting may be very fruitful.

Further research comprises implementation of fuzzy filters, association rules - in the spirit of [27] and some more advanced "interestingness" measures of the summaries. An appropriate combination of the querying and data mining functionality within the same interface poses also a challenge from the software engineering point of view.

References

1. Bosc P. and J. Kacprzyk (eds.), *Fuzziness in Database Management Systems.* Physica-Verlag, Heidelberg, 1995.
2. Bosc P. and O. Pivert, "Fuzzy querying in conventional databases". in: L.A. Zadeh and J. Kacprzyk (eds.): *Fuzzy Logic for the Management of Uncertainty.* Wiley, New York, 1992, 645-671.
3. Bosc P., L. Lietard and O. Pivert, "Quantified statements and database fuzzy querying, in P. Bosc and J. Kacprzyk (eds.): *Fuzziness in Database Management Systems.* Physica-Verlag, Heidelberg, 1995, 275-308.
4. Kacprzyk J. and S. Zadrozny, "FQUERY for Access: fuzzy querying for a Windows-based DBMS", in: P. Bosc and J. Kacprzyk (eds.) *Fuzziness in Database Management Systems*, Physica-Verlag, Heidelberg, 1995, 415 - 433.
5. Kacprzyk J. and S. Zadrozny, "Fuzzy queries in Microsoft Access v. 2.", in: D. Dubois, H. Prade and R.R. Yager (eds.): *Fuzzy Information Engineering - A Guided Tour of Applications*, Wiley, New York, 1997, 223 - 232.
6. Kacprzyk J. and S. Zadrozny, "Implementation of OWA operators in fuzzy querying for Microsoft Access", in: R.R. Yager and J. Kacprzyk (eds.) *The Ordered Weighted Averaging Operators: Theory and Applications*, Kluwer, Boston 1997, 293 - 306.
7. Kacprzyk J. and S. Zadrozny, "Flexible querying using fuzzy logic: An implementation for Microsoft Access", in: T. Andreasen, H. Christiansen and H.L. Larsen (eds.): *Flexible Query Answering Systems*, Kluwer, Boston, 1997, 247-275.
8. Kacprzyk J. and S. Zadrozny, "Data mining via linguistic summaries of data: An interactive approach", in T. Yamakawa and G. Matsumoto (eds.): *Methodologies for the Conception, Design and Application of Soft Computing* (Proceedings of IIZUKA'98), Iizuka, Japan, 1998, 668-671.
9. Kacprzyk J. and S. Zadrozny, "On sumarization of large datasets via a fuzzy-logic-based querying add-on to Microsoft Access", in: Intelligent Information Systems VII (Malbork, Poland), IPI PAN, Warsaw, 1998, 249-258.
10. Kacprzyk J. and Ziółkowski A., "Database queries with fuzzy linguistic quantifiers", *IEEE Transactions on Systems, Man and Cybernetics* SMC - 16, 1986, 474 - 479.
11. Petry F.E. *Fuzzy Databases: Principles and Applications.* Kluwer, Boston, 1996.
12. Vila M.A., Cubero J.C., Medina J.M. and Pons O., "Logic and fuzzy relational databases: a new language and a new definition", in P. Bosc and J. Kacprzyk (eds.) *Fuzziness in Database Management Systems.* Physica-Verlag, Heidleberg, 1994, 114 - 138.

13. Zemankova M. and J. Kacprzyk, "The roles of fuzzy logic and management of uncertainty in building intelligent information systems", *Journal of Intelligent Information Systems* 2, 1993, 311-317.
14. Yager R.R., "On linguistic summaries of data", in: G. Piatetsky-Shapiro and W.J. Frawley (eds.) *Knowledge Discovery in Databases*. AAAI Press/The MIT Press, Menlo Park, 1991, 347 - 363.
15. Rasmussen D. and R.R. Yager (1997) Fuzzy query language for hypothesis evaluation, in Andreasen T., H. Christiansen and H. L. Larsen (eds.) *Flexible Query Answering Systems*. Kluwer, Boston/Dordrecht/London, 23-43.
16. Agrawal R. and Srikant R., "Fast algorithms for mining association rules", in: Proceedings of the 20th International Conference on Very Large Databases, Santiago, Chile, 1994.
17. Zadeh L.A., "A computational approach to fuzzy quantifiers in natural languages", *Computers and Maths. with Appls.* 9, 1983, 149 - 184.
18. Zadeh L.A.,"A computational theory of dispositions", *International Journal of Intelligent Systems* 2, 1987, 39-64.
19. Weber S., "A general concept of fuzzy connectives, negations and implications based on t-norms and t-conorms", *Fuzzy Sets and Systems* 11, 1983, 115-134.
20. Anwar T.M., Beck H.W. and Navathe S.B., "Knowledge mining by imprecise querying: A classification based system", in: Proceedings of the International Conference on Data Engineering, Tampa, USA, 1992, 622-630.
21. George R. and R. Srikanth, "Data summarization using genetic algorithms and fuzzy logic", in: F. Herrera and J.L. Verdegay (eds.): *Genetic Algorithms and Soft Computing*. Physica-Verlag, Heidelberg and New York, 1996, 599 - 611.
22. Mannila H., Toivonen H. and Verkamo A.I., "Efficient algorithms for discovering association rules", in: U.M. Fayyad and R. Uthurusamy (eds.) Proceedings of the AAAI Workshop on Knowledge Discovery in Databases, Seattle, USA, 1994, 181-192.
23. Srikant R. and Agrawal R., "Mining generalized association rules", in: Proceedings of the 21st International Conference on Very Large Databases", Zurich, Switzerland, 1995.
24. Srikant R. and Agrawal R., "Mining quantitative association rules in large relational tables", in: Proceedings of the ACM-SIGMOD 1996 Conference on Management of Data, Montreal, Canada, 1996.
25. Srikant R., Vu Q. and R. Agrawal, "Mining association rules with item constraints", in: Proceedings of the 3rd International Conference on Knowledge Discovery in Databases and Data Mining, Newport Beach, USA, 1997.
26. Miller R.J. and Yang Y., "Association rules over interval data". in: Proceedings of the ACM SIGMOD International Conference on the Management of Data, Tucson, USA, 1997, 452-461.
27. Lee J.-H. and Lee-Kwang H., "An extension of association rules using fuzzy sets", in: Proceedings of the Seventh IFSA World Congress, 1997, Prague, Czech Republic. Vol. 1, 399-402.

Part III.

Geographic Information Systems Dealing with Uncertainty

Uncertainty in Distributed and Interoperable Spatial Information Systems

M. Cobb[1], H. Foley[2], F. Petry[3], K. Shaw[4]

[1] Department of Computer Science and Statistics, University of Southern Mississippi, Hattiesburg, MS 39406-5106
[2] Department of Computer Science, Xavier University, New Orleans, LA 70125
[3] Department of Electrical Engineering and Computer Science, Tulane University, New Orleans, LA 70118
[4] Naval Research Laboratory, Stennis Space Center, MS 39529

Abstract. Many facets of spatial data representation inherently involve issues of accuracy and uncertainty. This problem is greatly magnified when considering the integration of spatial data from different sources, such as in a distributed or interoperable environment. The general concept of schema merging involves the resolution of incompatibilities as in a distributed environment. These may be either structural or semantic in nature. Structural incompatibilities involve those, for example, in which attributes for representing the same values are defined differently. Semantic incompatibilities, however, represent those cases in which similarly defined attributes have different meanings or values. For example, an attribute of WIDTH for a road in one database may include the width of associated access lanes, while in another database it may be only the main driveable portion of the road. Such semantic issues are much more difficult to resolve, as they require a deeper understanding of the data. We will survey the issues as discussed above for spatial data in such environments and describe several approaches for different aspects of the data using fuzzy set techniques to deal with the incompatibilities.

Keywords. fuzzy GIS, conflation, uncertainty, spatial databases

1 Introduction to Spatial Data Uncertainty

The database field is currently extending its paradigm from a mostly relational approach towards an object-oriented direction. It is not clear yet what forms these object-oriented extensions will take. Indeed, there is little agreement upon which features are needed, since this approach did not spring from a single source as the relational model did from Codd's research [1]. Likewise, the scope of database capabilities is expanding to encompass multimedia, including text, pictures and sound, particularly in a distributed environment. This is of significance in supporting complex applications such as CAD/CAM design systems and geographical information systems (GIS). Such a range of database approaches represents part of a broad spectrum of information systems encompassing even the processing of large text-type files conventionally construed as the realm of information storage and retrieval systems.

The importance of representations of uncertainty in databases is increasing as more complex applications such as CAD/CAM and GIS are being undertaken in object-oriented and multi-media databases. In this paper we consider a number of issues related to uncertainty in spatial data representations and GIS, and the impact on a distributed system in dealing with them. The relational model has been the dominant database model for a considerable period, and so it was naturally used by researchers to introduce fuzzy set theory into databases. Much of the work in the area has been in extending the basic model and query languages to permit the representation and retrieval of imprecise data. A number of related issues such as functional dependencies, security, implementation considerations and others have also been investigated.

Two major approaches have been proposed for the introduction of fuzziness in the relational model. The first one uses the principle of replacing the ordinary equivalence among domain values by measures of nearness such as similarity relationships [2], proximity relationships [3], and distinguishability functions [4]. The second major effort has involved a variety of approaches that directly use possibility distributions for attribute values [5, 6, 7]. There have also been some mixed models combining these approaches [8, 9]. These approaches have also been extended to object-oriented databases [10]. We can see that variations of all of these approaches are of use in modeling spatial and geographic data, and some will be described shortly.

The need to handle imprecise and uncertain information concerning spatial data has been widely recognized in recent years (e.g., [11]), particularly in the field of GIS. GIS is a rather general term for a number of approaches to the management of cartographic and spatial information. Most definitions of a geographic information system [12, 13] describe it as an organized collection of software systems and geographic data able to represent, store and provide access for all forms of geographically referenced information. At the heart of a GIS is a spatial database. The spatial information describes the location and shape of geographic features in terms of points, lines and areas.

There has been a strong demand to provide approaches that deal with inaccuracy and uncertainty in GIS. The issue of spatial database accuracy has been viewed as critical to the successful implementation and long-term viability of GIS technology [11]. The value of a GIS as a decision-making tool is dependent on the ability of decision-makers to evaluate the reliability of the information on which their decisions are based. Users of geographic information system technology must therefore be able to assess the nature and degree of error in spatial databases, track this error through GIS operations, and estimate accuracy for both tabular and graphic output products. There is a variety of aspects of potential errors in GIS encompassed by the general term "accuracy." However, here we focus on those aspects that lend themselves to modeling by fuzzy set techniques.

1.1 Significance of Uncertainty Modeling for Spatial Data Systems

There have been a number of recent indications of the importance of uncertainty modeling in spatial data. Two in particular are of most significance. First, the

National Imagery and Mapping Agency of the United States (NIMA), announced for fiscal year 1997 a new program of University Research Initiatives. One of the major topics is uncertainty in geospatial information representation, analysis and decision support. The following are some of the main aspects:

i. *Elements of Uncertainty*: Geospatial information is extremely complex and includes several aspects that may have associated uncertainties. These include location, relationships and typologies. They requested proposals to identify and describe all aspects of uncertainty associated with geospatial information.
ii. *Models for Uncertainty:* The goal here was to develop extensions to existing geospatial data models that accommodate the elements of uncertainty.
iii. *Propagation of Uncertainty:* For this aspect of the proposed efforts, they requested development of algorithms for determining how uncertainty is propagated through the fusion and analysis of geospatial information.

Secondly, the University Consortium for Geographic Information Science (UCGIS) has published a major position paper [14] of research priorities for geographic information science. In this, they state that the uncertainty information associated with a geographic data set can be conceived as a map depicting varying degrees of uncertainty associated with each of the features or phenomena represented in the data set, and potentially separable into three components: uncertainty in the typological attributes (describing the type of a geographic feature), uncertainty in the locational attributes, and uncertainty in spatial dependence (the spatial relationship with other features).

Uncertainty is seen as appearing in every part of the geographic data life cycle: data collection, data representation, data analyses, and results. The data that passes through the stages of observation to eventual archiving may be handled by a variety of individuals/organizations, each of whom may provide their own distinct interpretations to the data. So, the uncertainty is mostly a function of the relationship between the data and the user, i.e., a measure of the difference between the data and the meaning attached to the data by its current user. The UCGIS emphasized that research was needed in studying in detail the sources of uncertainty in geographic data and the specific propagation processes of this uncertainty through GIS-based data analyses, in developing techniques for reducing, quantifying, and visualizing uncertainty in geographic data, and for analyzing and predicting the propagation of this uncertainty through GIS-based data analyses.

1.2 Sources of Imprecision in Spatial Data

There is a variety of sources of imprecision in geographic information systems that are manifested as several types of uncertainty: (1) Uncertainty due to variability or error; (2) Imprecision due to vagueness; and (3) Incompleteness due to inadequate sampling frequency for missing variables [15]. Both uncertainty of interpretation and inherent ambiguity are illustrated by the labeling of data such as that obtained from Landsat images. The images are initially processed by unsupervised classifications to obtain image classes and then the results are subjectively assigned land cover or resource class labels by a human interpreter. This is an inherently subjective task in which the interpreter attempts to match objectively derived image classes with linguistic concepts that are represented in the mind of the interpreter. It is not surprising that there is variation in the interpretation of the same data among interpreters. This is particularly troublesome when the result is stored in a database, because at this point an inherently imprecise concept requires a specific representation. As a result, uncertainty modeling in GIS has been used for the classification of land according to soil type, vegetation cover, and land use (e.g., [16]). Uncertainty representations in both probabilistic [17] and fuzzy [18] frameworks [19] have been utilized, and Fisher [20], [21] has striven to elucidate the fundamental differences between them in this context. In applications involving remotely sensed information and typical multiple sources of information used to formulate geographical data, the problems of imprecision and uncertainty are of even more concern [22].

Many operations are applied to spatial data under the assumption that features, attributes and their relationships have been specified a priori in a precise and exact manner. However, this assumption is generally not justifiable, since inexactness is almost invariably present in spatial data. Inexactness exists in the positions of features and the assignment of attribute values, and may be introduced at various stages of data compilation and database development. Moreover, inexactness may be propagated through GIS operations to appear in modified form on tabular and graphic output products. Inexactness is often inadvertent, as in the case of measurement error or imprecision in taxonomic definitions, but may also be intentional since generalization methods are frequently applied to enhance cartographic fidelity.

Models of uncertainty have been proposed for GIS information that incorporate ideas from natural language processing, the value of information concept, non-monotonic logic and fuzzy set, evidential and probability theory. For example in [23] there are reviews of four models of uncertainty based on probability theory, Shafer's theory of evidence, fuzzy set theory and non-monotonic logic. Each model is shown as appropriate for a different type of inexactness in spatial data. Inexactness is classified as arising primarily from three sources. "Randomness" may occur when an observation can assume a range of values. "'Vagueness" may result from imprecision in taxonomic definitions. "Incompleteness of evidence" may occur when sampling has been applied, there are missing values, or surrogate variables have been employed.

It is now clear that uncertainty concepts must be part of the apparatus of GIS [24]. Limited positional accuracy is found in maps because of the process of map production and the inherent limitations of Earth measurement systems. Positional accuracies better than about 1 part in 10^4 or 10^5 are uncommon in paper maps [11], but much greater precision is available in computational systems. Many designers represent positions using double precision (1 part in 10^{14}), and at this accuracy, a map of the entire Earth would be capable of recording accurately the positions of large protein molecules.

Scale effects have also been considered for uncertainty modeling approaches. Maps are always generalized, to a degree that depends largely on their representative fraction or scale, or the ratio of distance on the map to distance on the ground. A map at scale 1:24,000 is less generalized than one at 1:100,000. Often, data are not available at the scale required by an application, but only at a coarser scale. As an example, to obtain a sufficiently accurate assessment of a given area's suitability for development may require data at the level of generalization corresponding to 1:24,000, but the only available digital data may be at a scale of 1:100,000. In this situation, uncertainty introduced into the results of analysis when too-coarse data are used must be considered. Ehlschlaeger et al. [25] and others have developed methods for simulating the information that is not available due to excessive generalization, based on geostatistical models calibrated in areas where both data are available at both scales.

2 Application of Fuzzy Models in GIS

2.1 Spatial Data and Attribute Modeling

Robinson [26, 27, 28] performed some of the earliest research on fuzzy data models for geographic information. He has considered several models as appropriate for this situation—the two early fuzzy database approaches using simple membership values in relations by Giardina [29] and Baldwin [30], and a similarity-based approach [2]. In modeling a situation in which both the data and relationships are imprecise, he assesses that this situation entails imprecision intrinsic to natural language which is possibilistic in nature. A possibilistic relational (PRUF) model was chosen as providing a means of facilitating approximate machine inference [31]. In the PRUF model, queries and propositions are processed by identifying constraints induced by the query or proposition, performing tests on each constraint and then aggregating the individual test results to yield an overall test score. Consider a proposition stating that a specified location is on gentle slopes and is near a certain city. The constraints induced by the proposition, "gentle" and "near," are tested using a possibility distribution yielding test results indicating the degree to which the specified location satisfies each constraint. The two test results are then aggregated to produce an overall test score indicating the degree to which the proposition is satisfied.

A number of more recent papers have also appeared that use fuzziness in modeling imprecision in spatial data. Cross [32] presents various issues in the area of fuzzy object-oriented databases and design and how these relate to topics involving GIS. Usery [33] developed an object-based model in which objects are a direct representation of the geographic entities rather than geometric elements such as point, line, and area. This design is referred to as a feature-based geographic information system (FBGIS) since the term *feature* encompasses both the geographical entity and its object representation. The approach taken is to model the spatial dimension of geographic features using fuzzy sets to define the spatial extent of the thematic dimension.

The focus is on features with undetermined boundaries which are themselves fuzzy and, regardless of the accuracy of measurement, remain fuzzy based on the concept of the feature. Examples include hills, valleys, wetlands, and many other geographic features that cannot be rigorously bounded by a mathematical line. The model of a fuzzy feature follows that of Leung [34] using the concepts of core and boundary.

For the hill case, the core defines the prototype elevation values, i.e., all values that are unquestionably a part of the hill. The periphery is defined by the boundary and is less representative of the hill. As the complexity of the feature increases, the complexity of the fuzzy set membership function also increases. Once fuzzy set membership is defined, one can define fuzzy operations in a GIS environment. Using established fuzzy set operations such as intersection, union, and complement, a fuzzy overlay operator can be developed. Fuzzy buffering, fuzzy overlay, including intersection, union, and complement, and fuzzy boundary operators are used as have been developed by Katinsky [35].

Another paper in the same volume by Sarjakoski [36] emphasized issues related to identity of geographical objects. This issue arises in the practical context of identifying objects in a complex and dynamic environment; for example, the question of enumerating how many lakes, islands and rivers are in Finland.

Existence is a necessary property of an object. An object also has an identity. This is the key issue when distinguishing between field-(in the mathematical sense) and object-oriented approaches in describing geographic reality [37, 38]. There is no identity related to fields as such. Objects can be, and very often are, associated with regions in a field; thus, fields are indirectly connected with the concept of identity.

Many of the above concepts are closely interwoven. Take the triple existence-extent-identity, for example. It is difficult, or even impossible, to speak of one without another. According to our contemporary understanding of the universe, geographic objects can only exist confined in four-dimensional space-time. When there appears to be fuzziness, it occurs in many of the properties of an object, not in one alone. Some of the extreme cases are:

Fuzziness only in 3D space: Fuzziness may be limited to the spatial extent of a geographical object, i.e., the boundary of the object is ill-defined, even though there is no doubt about its identity and temporal extent.

Fuzziness only in time: The temporal extent of a geographic object may be ill-defined, even though the spatial extent is well-defined, i.e., the spatial boundary is sharp.

Fuzziness in identity: This is closely related to fuzziness in spatial and temporal extent. The issue here is whether an object has enough identity to be an individual. With dynamic objects (objects that change in time), the concept is linked to the problem of the extent to which an object can change and still be the same object.

Fuzziness in class definition: Class definitions are sometimes fuzzy and overlap, making it impossible to assign an object to only one class.

Fuzziness in class membership: Even when there are well-defined (axiom- or definition-based) classes, the characteristics of an object may still be such that we cannot be quite sure to which class it belongs.

In an actual survey, heuristic and rather informal classification rules were used to define lakes, rivers, and islands. Much criteria cited involved linguistic descriptions and modifiers.

In principle and by definition, a geographic object will always have an identity and therefore a crisp existence. This implies that a geographic object can always have an individual name. Sarjakoski's case study demonstrates that, in reality, it is often very difficult to state whether or not something can be modeled as an individual object. The objects on this conceptual borderline can be said to be fuzzy in existence. In other words, it is not clear whether the object exists at all: fuzzy existence is always strongly related to fuzziness in extent, whether spatial or temporal. The task of counting the number of geographic objects deals only with their existence; it is in principle unnecessary to know their extent. Nevertheless, because the objects are located in the same geographic field, their spatial extent must be determined in a consistent way.

A recent system developed by Robinson is the Knowledge Based Land Information Manager and Simulator (KBLIMS) [39] which can provide a perspective on fuzzy sets as a basis for managing certain kinds of uncertainty in a geographic information system.

Several domains were considered for the use of fuzzy sets to model uncertainty in a comprehensive, intelligent geographical information system application such as KBLIMS. On the terrain analysis level, uncertainty representation is required for the hydro ecological simulation system as well as the intelligent query portion of the system. The uncertainty model of Davis and Keller [40] shows how to incorporate fuzzy metadata concepts with Monte Carlo simulation in a system that can use the uncertainty information to provide an assessment of the validity of the model output. The most direct approach for the integration of fuzzy sets at the terrain level in KBLIMS has the uncertainty within the process of terrain analysis so that the output is in a crisp form. Otherwise, the objects formed from the terrain analysis would need a capability to represent and interpret uncertainty at the object model level.

The object model level of KBLIMS is where management of uncertainty using fuzzy sets can be made most explicit and thoroughly integrated into the system. Incorporation of fuzzy instances, objects, and classes depends greatly upon the nature of the data output from the terrain analysis and the needs/capabilities of the query model. In this way, the traditionally separate fields of representation and spatial analysis [41] become integrated. A working object model in KBLIMS requires implementation of the behavior of the fuzzy objects, relations, or classes. Often, this would be the same as many spatial analysis functions, except that some spatial analysis functions would be incorporated into portions of the query model.

KBLIMS and other sophisticated GIS applications have not been initially designed with the use of fuzzy sets in mind for uncertainty management. There are portions of the system which may or may not use fuzzy sets to address very specific problems such as model self-evaluation, allowing the system to enhance its ability to manage uncertainty generated by its own activities. However, as the application domain evolves and experience with formalizing that domain increases, some significant kinds of uncertainty may be specified and ultimately managed with fuzzy sets. It is also evident that in GIS applications such as KBLIMS, that simply arriving at membership values to describe variations in soils and their properties over space is only the beginning of the development of a process for formalizing and managing uncertainty that will spread through the system, requiring changes at the object model level.

2.2 Spatial Relationship Modeling

The manner in which spatial data is modeled affects important aspects of its use, including querying capabilities and relationship inferences. Especially important for spatial data is the ability to model relationships between spatial features. Such relationships can provide a significant resource for processing spatial queries.

A spatial data model that provides a representation for storing information concerning binary relationships between two-dimensional objects is described in previous work [42, 43, 44]. The binary relationships incorporated include both qualitative topological and directional relationships. In this approach, "qualitative topological relationships" include both those relationships based on topological invariants, e.g., Egenhofer [45], and a more intuitive, less formal set of relationships based on various other properties. The model represents a compromise between storage and performance issues by explicitly storing information that can easily be used for inference purposes by a fuzzy query framework, such as that given in [44, 46].

Minimum Bounding Rectangles (MBRs) are used as the basis for object representation. This approach, as well as that of Nabil [47] Sharma [48] and Clementini [49], relies upon the use of MBRs as approximations of the geometry of spatial objects. The use of MBRs in geographic databases is widely practiced as an efficient way of locating and accessing objects in space [49]. In addition, numerous spatial data structures and indexing techniques have been developed that exploit the computationally efficient representation of spatial objects through the use of MBRs [50, 51].

This model also supports fuzzy relationship representation and querying of specific aspects of direction and qualitative topology. For direction, it is used to answer queries related to the *degree* to which one object lies in a particular direction with respect to a second object. The determination of the degree is made through calculations that utilize area and axis ratios of the objects involved. The resulting quantitative value is then mapped to a range that corresponds to a term known as a *directional qualifier.* Directional qualifiers can be used in queries to determine, for example, that *object A is mostly west of object* B, or that *object A is somewhat southeast of object B.*

The need for such directional qualifiers is evident when considering the difficulty of determining a precise directional relationship for 2-D objects. The areal extent of 2-D objects often results in the possible application of more than one directional relationship. Often, centroids are used to provide a single point of reference for determining orientation of 2-D objects, (e.g., Chang [52]). However, some information loss is inherent in the centroid method. This model preserves all of the directional relationships that exist between two 2-D objects, along with a representation indicating the degree to which the objects are analyzed to participate in each of the relationships. This method of qualifying directional relationships more closely models the way in which humans naturally process and communicate such information.

Qualitative topological relationships such as *overlaps, surrounded-by, disjoint,* etc., are similarly modeled so that the degree to which each of the objects participates in the relationship is stored for subsequent fuzzy querying. Determination of this information utilizes various combinations of area ratios, ranges of which are mapped to a set of linguistic qualifiers such as *some, most, all,* etc., in a manner analogous to that used for directional relationships. These qualifiers allow users to distinguish the various cases of the same relationship. For example, given that A overlaps B, does *all* of A overlap *some* of B, or does *little* of A overlap *most* of B? This type of distinction is frequently made during human-to-human interaction, but little support for it has been incorporated into computer data models.

The definition of the relationships is based on an extension of Allen's temporal relations [53] to the spatial domain. This is done by allowing each of Allen's thirteen relations to represent the interaction of two-dimensional objects in terms of an x and y relationship component. The resulting set of relationships is then used for defining topological and directional relationship terminology. A data structure called an abstract spatial graph (ASG) is defined for the binary relationships. The ASG maintains all necessary information regarding topology and direction, and provides the basis for processing of fuzzy topological and directional queries.

The ASG model was specifically designed to store explicit information regarding fuzzy topological and directional relationship information for pairs of objects. This correlates well with the observation in [45] that topological information should be considered first-class information within the realm of spatial databases.

Briefly, an ASG is a directional polar graph with three types of nodes: (1) nodes representing portions, or object sub-groups of the first object participating in the relationship, (2) nodes representing object sub-groups of the second object, and (3) nodes representing overlapping areas and/or reference areas. For the graphical representation of an ASG, the various types of nodes are differentiated through color and symbol style. As stated earlier, one of the basic assumptions for the data model and subsequent query framework is that objects, or features, are enclosed within MBRs. Reference areas for relationship calculations are defined as being either one of the object sub-groups, a specially defined rectangular area, or a line segment--depending upon the general categorization of the relationship involved.

Each node of an ASG has a pair of associated weights: an area weight and a total node weight, that provides information to support fuzzy querying. Specifically, these weights are used to define fuzzy qualifiers that answer queries such as, "To what degree is area A east of area B?" or, "How much of area A surrounds area B?." Total node weights support queries of the first type, while area weights support queries of the second type. Ranges of weights are used to define qualitative terms useful for describing these concepts, such as "most," "some," "slightly," or "mostly." The manner in which the weights are computed is described in [43].

The ASG data model supplies the infrastructure upon which a fuzzy query system is built. This system, topological query framework (TQF), includes a grammar in Backus-Naur Form (BNF), data structures and query processing strategies. Fuzzy relationship querying follows naturally from the ASG model due to its support for direct binary relationship representation and weighting system. Previous work [46] has shown how TQF can be integrated with several existing spatial query languages to provide fuzzy querying extensions.

3 Conflation

Conflation is typically regarded as the combination of information from two digital maps to produce a third map that is better than either of its component sources. The history of map conflation goes back to the early to mid-1980's. The first clear development and application of an automated conflation process occurred during a joint United States Geological Survey (USGS)-Bureau of the Census project designed to consolidate the agencies' respective digital map files of U.S. metropolitan areas [54]. The implementation of a computerized system for this task provided an essential foundation for much of the theory and many of the techniques used today. Since that time, others, including commercial GIS vendors, have implemented conflation tools within their applications. For an example of commercial work on conflation, see [55].

Conflation of maps typically is needed because: (1) users wish to update their mapping information without losing legacy data which may not be included in the new information; (2) one map source may be more accurate with respect to, e.g., positional or attribute information; or (3) one map source contains information missing in another, such as additional features, feature attributes or even entire coverages. Motivation for this work includes all three of these reasons and represents an effort to more fully utilize associated non-spatial data as well as spatial properties in an “intelligent” type of system that more closely mirrors the way that human experts would manually conflate mapping data.

Conflation can, in brief, be viewed as a multi-step, iterative process that involves feature matching, positional re-alignment of component maps and attribute deconfliction of positively identified feature matches. Feature matching, simply and perhaps somewhat obviously stated, involves the identification of features from different maps as being representations of the same geographic entity. Positional alignment is a mathematical procedure in which previously identified matching features are brought into spatial agreement, while deconfliction is a process in which contradictions in a matching pair's attributes and/or values are resolved. Positional alignment and deconfliction are both steps that are performed after a positive match has been determined. As such, it is easy to see that accurate feature matching results are essential to the overall quality of the resulting conflated map. Because of this dependency, our work thus far has concentrated solely on feature matching aspects of conflation. Our motivation for this work includes an effort to make individual geographic features "intelligent" enough to know when and how to conflate themselves in a distributed environment.

3.1 Conflation and Uncertainty

The assessment of feature match criteria is a process in which evidence must be evaluated and weighed and a conclusion drawn—not one in which equivalence can be unambiguously determined. For example, fuzzy concepts such as "closeness" of two features and "similarity" of attributes and feature groupings are essential for determining equivalence.

In fact, feature matching can be considered as a type of classification problem. That is, we are trying to determine whether one feature belongs to the same "class" as another; in this case the class consists of members believed to be equivalent. This type of problem can be handled through theories of evidential reasoning or uncertainty, such as fuzzy logic [57] or Dempster-Shafer theory [58]. These theories attempt to provide likelihood measures for questions based on available, though not necessarily conclusive, evidence. For example, in feature matching, the question we must consider is, "Based on the available evidence, what is the likelihood (or probability) that feature A from map coverage 1 represents the same entity as feature B from map coverage 2?"

Dempster-Shafer theory is a method of inexact reasoning based on the modeling of uncertainty as a range of probabilities. It provides representations for the degree of belief in evidence (belief that supports a hypothesis), as well as the degree of disbelief in evidence (belief that refutes a hypothesis) and the nonbelief (neither belief nor disbelief). That is, it not only supplies results regarding the belief that two features are the same, but it also provides a representation of how much it is disbelieved that they are the same, along with the degree to which there is no evidence either way.

The approach described here for feature matching draws from aspects of both fuzzy set logic and evidential reasoning. In particular, the assignment of matching scores for linguistic attributes is directly motivated by work in modeling linguistic variables by the use of fuzzy sets. For example, we need to be able to determine the semantic similarity of an attribute such as **road surface type** which may have a value of 'hard' for one feature and 'asphalt' for the potential matching feature. Obviously, the semantics of the two are not strictly equivalent, but neither are they contradictory. Likewise, the idea of combining scores from the different components of feature matching to arrive at a single matching score is very similar to techniques for the combination of evidence used in evidential reasoning.

Previous work [59, 60, 61] has concentrated on developing an uncertainty model for feature matching. Human cartographic experts are skilled in making mapping decisions based on conflicting information, and it is this skill at reasoning with uncertain information that we attempted to mimic in this system. Geographic features represented in different physical databases are very unlikely to have completely identical information due to differences in collection instruments, human interpretation, processing techniques and varying representations of the data. Therefore, a conflation model must be able to determine when such differences are truly indicative of non-matching features, and when such differences could be attributed to one of the causes identified above.

The technique developed in [59] is able to accommodate cases where values for corresponding attributes differ, as well as cases where the attribute sets themselves differ. For implementation, each feature is considered as a set of attribute-value pairs:

$((a_{11}, v_{11}), (a_{12}, v_{12}), \ldots, (a_{1n}, v_{1n}))$
$((a_{21}, v_{21}), (a_{22}, v_{22}), \ldots, (a_{2m}, v_{2m}))$
...

From this representation, a degree of matching similarity is determined. A different approach is used for different attribute value domains. For numeric domains, a membership matching function is used, while a similarity table is used for linguistic domains. Our approach to linguistic attribute matching is to establish a similarity value *s* (in the range [0, 1]) for each pair of elements of the domain. This value is determined from the semantics of the domain and the linguistic terms. The characteristics of the similarity function *s* are:

$$s_A(x, y) = s_A(y, x) \qquad \underline{\text{symmetric}}$$

for all values x, y ∈ domain of attribute A. For well-defined values of the domain (e.g. not "unknown" or "other")

$$s_A(x, y) = 1 \qquad \underline{\text{reflexive}}$$

where x is a well-defined value.

As an example, we consider the **Railroad** attribute *RRA* (Railroad Power Source), which is restricted to the values (0—Unknown, 1—Electrified track, 3—Overhead electrified, 4—Non-electrified, 999—Other). The similarity table for *RRA* is shown in table 1. Since linguistic similarity is symmetric, we need only show the lower triangular values in the table. Note that since 1, 3, and 4 are well-defined linguistic terms, they are shown with the reflexive value of 1 on the diagonal, e.g. $s_{RRA}(3, 3) = 1$. However, since 0 and 999 are non-specific categorical values for this particular domain, their diagonal values were determined to be less than 1, reflecting the potential lack of exact matching for such linguistic terms.

Table 1. Similarity table for attribute *RRA*.

RRA	0	1	3	4	999
0	.2				
1	.2	1			
3	.2	.6	1		
4	.2	.1	.1	1	
999	.2	.2	.2	.2	.2

Because most features have more than one attribute, we must also consider semantic interrelationships among the attributes in determining matching features. These are represented as rules in the expert system which return associated weights. These weights are used either to add credence to the hypothesis of matching features, or to weaken it. As an example, consider the **Railroad** attributes *LTN* (Number of tracks) and *RRC* (Railroad category). The rule for the relationship between the two attributes is expressed as:

IF ((RR1.ltn = 3 and RR2.ltn = 2) and
(RR1.rrc = 16 and RR2.rrc = 16))
THEN $w_{rra} \leftarrow 1.0$
$w_{ltn} \leftarrow 0.5$,

where RR1 represents the first **Railroad** object and RR2 represents the second. This rule illustrates a conflict in the values of the length attribute of the two features. We see from the example that the resulting weight for *LTN* is weakened, giving it less influence than that of *RRA* in the combined matching score.

A composite matching score is then computed from the combination of the expert system weights and the similarity table values. This score is given as:

$$MS_{i,j} = \left(\sum_{k=1}^{N} [simA_k(F_i, F_j) \times ESW_{A_k}]\right) / N$$

where
A_k = kth attribute in both F_i and F_j, where $0 \le k \le N$.
N = number of attributes that describe both F_i and F_j.
ESW_{A_K} = weight associated with A_k computed by the expert system.

3.2 Geometric Analysis

The second part of this work consists of a technique for qualitatively matching linear features based on shape similarity. This broadens the basis for feature matching by adding a spatial component to the non-spatial attribute-matching algorithm presented above.

The first step involves a pre-processing of the features to bring them into a standard position for the shape similarity analysis. Translation, rotation, and scaling transformations are applied in sequence to accomplish this positioning. Translation is performed by moving the features' starting nodes to the origin of planar axes. Then, the end nodes are rotated to the x-axis. Finally, the features are scaled so that the starting and ending nodes are equal to [0,0] and [1,0], respectively. These transformations allow us to more reliably compare linear features of different sizes. Figure 1 shows the results of standardizing two linear features through this process.

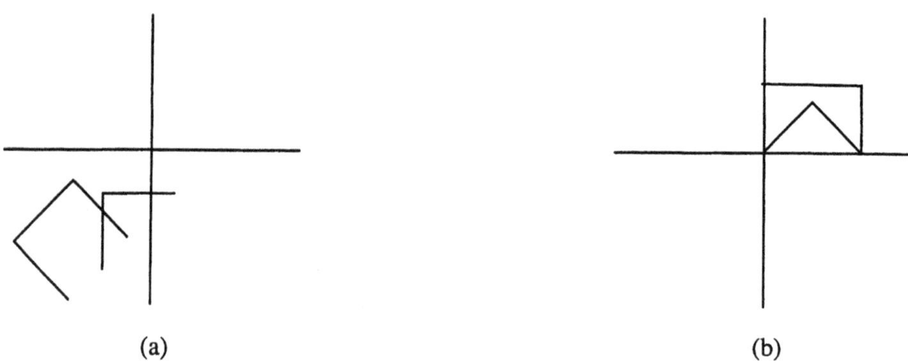

Figure 1. Linear features (a) before and (b) after standardization process.

After the features have been standardized, a process called *node merging* takes place [54]. This requires all nodes from one feature to be mapped onto the other feature, and vice versa. This is performed on the basis of a node's distance from the starting node of the feature. Consider two features, F1 and F2, comprised of nodes (n_{11}, n_{12}, ..., n_{1j}) and (n_{21}, n_{22}, ..., n_{2k}), respectively, and where the associated ratios are given as (r_{11}, r_{12}, ..., r_{1j}) and (r_{21}, r_{22}, ..., r_{2k}) such that $r_{11} = r_{21} = 0.0$ and $r_{1j} = r_{2k} = 1.0$, where r_i is the ratio of the distance from start to node n_i to the overall length of the linear feature. Then, new features F1 and F2 are created that each have the same number of nodes such that each node is the result of merging the features' nodes based upon their ratios.

Given that the features are in standard position and the nodes have been merged, we begin the last phase in assessing shape similarity. This involves determining a normalized distance measure between corresponding nodes of the two features, given as

$$SHsim = \left(\sum_{i=1}^{j+k-2} Dist(n_{1i}, n_{2i}) / (D_1 + D_2)\right)$$

where j + k - 2 is the number of distances to be computed, and D_i is the total length of F_i. This implies that features with a higher shape similarity measure are more likely candidates for matching features. It is based on Frechet's measure of distance, L_2 - *Distance*, between polygonal arcs [55]. Suppose that we have two functions

$$P : [0,1] \rightarrow R^2$$
$$Q : [0,1] \rightarrow R^2$$

where $(p_0, p_1, \ldots, p_n)$ and $(q_0, q_1, \ldots, q_m)$ are ordered point sets for P and Q, respectively. Then the set of ratios used for computing the distance between P and Q at all of the r_i's is computed as described for node merging above. Thus, for Frechet's measure, the following is used.

$$L_2 = \left(\int_0^1 \| P(r) - Q(r) \|^2 dr \right)^{1/2}$$
$$= \left(\sum_{i=1}^{n+m} \int_{r_{i-1}}^{r_i} \| P(r) - Q(r) \|^2 dr \right)^{1/2}$$

Now, given that (x_{Pi}, y_{Pi}) and (x_{Qi}, y_{Qi}) are the coordinates of $P(r_i)$ and $Q(r_i)$, respectively, and $\Delta_{x,i} = x_{P_i} - x_{Q_i}$ and $\Delta_{y,i} = y_{P_i} - y_{Qi}$ are the measured separations of the linear features at the bending points of either of the lines, then the closed form of the equation for L_2 - *Distance* is represented by

$$\left\{ 1/3 \sum_{i=1}^{n+m} \left((r_i - r_{i-1}) \left(\Delta_{x,i-1}^2 + \Delta_{x,i-1}\Delta_{x,i} + \Delta_{x,i}^2 + \Delta_{y,i-1}^2 + \Delta_{y,i-1}\Delta_{y,i} + \Delta_{y,i}^2 \right) \right) \right\}^{1/2}$$

Of course, the mathematical exposition presented here presents the objective portion of feature matching. The subjective portion represented by the expert system is much more difficult to capture. In response to this problem, Foley et al. [60, 61] present a web-based knowledge acquisition tool for capturing expert knowledge needed for conflation.

4 Distributed Spatial Systems

In recent years, the trend for most types of information systems has been a move to a more loosely coupled, distributed nature. The maturity of client-server architectures and software, as well as the virtual explosion of web-based systems, has demonstrated the enormous advantages of distributed systems. Furthermore, the advent of successful middleware technology such as the CORBA 2.0 standard and corresponding vendor implementations has drastically reduced barriers to communication and data sharing among heterogeneous software and hardware systems.

The interest in distributed GIS is no less than that of general information systems; however, the uniqueness of the nature of spatial data makes the issue of true *interoperability* of GIS a major research concern. Evidence of this is abundant in the literature, and is also illustrated by initiatives such as the Open Geodata Interoperability Specification (OGIS) work by the Open GIS Consortium, Inc., as well as UCGIS priority research panels [14] on "Interoperability of Geographic Information" and "Spatial Data Acquisition and Integration."

4.1 Uncertainty and Conflation for Distributed Data

Obviously, issues of uncertainty that apply to conflation within a single system—such as those for feature matching—are also applicable to conflation in a distributed environment. However, we believe additional factors related to the general topic of distributed databases increase the scope of uncertainty that must be considered in this context. As background, we can draw from the abundance of past and ongoing research in the realm of schema merging for conventional (i.e., relational) distributed heterogeneous databases. An example of work in this area includes [62].

The general concept of schema merging involves resolution of incompatibilities in metadata. These incompatibilities may be either *structural* or *semantic* in nature. Structural incompatibilities involve those, for example, in which attributes for representing the same values are defined differently. These may include different names for the attributes, or different domains for their associated values, e.g., float vs. integer. Semantic incompatibilities, on the other hand, represent those cases in which similarly defined attributes have different meanings or values. For example, an attribute of width for a road in one database may include the width of the road plus any associated right-of-ways, while the same attribute name in another database may only imply the width of the paved/driveable portion of the road. Semantic incompatibilities are much more difficult to handle automatically, as they necessarily imply a deeper understanding of the data.

It is clear that these issues are very similar to ones that must be faced in performing conflation in distributed spatial databases. In particular, semantic schema integration and the feature-matching phase of conflation require similar levels of knowledge regarding the meanings that various data are intended to convey. Semantic knowledge is inherently uncertain, as interpretations of even the most seemingly unambiguous words and phrases vary among individuals. Similarly, structural differences in spatial data representation for like features are to be expected in any distributed system comprised of heterogeneous data sources. It is evident from this discussion that conflation in a distributed environment can be viewed as a specific application of issues related to uncertainty in schema merging.

In keeping with generally accepted principles of distributed (as well as non-distributed) database design, the model presented in [63] is completely independent of physical concerns such as data partitioning. However, the general issue of considering conflation within a distributed system versus a standalone system does impact design decisions, even at an abstract logical level. The primary issue considered in this work is centralization versus distributed control of conflation events.

4.2 Object-Oriented Modeling Approach

The logical design for this approach is based on an object-oriented (OO) model. The OO paradigm is well accepted as the prevailing method for the representation and manipulation of complex data such as geographical information. Within an OO framework, one is able to define models of real-world data in ways analogous to those in which we intuitively perceive and interpret those data. As a general introduction to the subject, an *object* is a collection of data (state) and methods (behavior) that represents the properties and processes of a real-world entity, such as the Chesapeake Bay Bridge or the Mississippi River. A *class* is a template for creating new objects that share common properties. For example, there may be a bridge class that captures generic information that every *instance* of that class (an instantiation) should contain. This would most likely include data such as length, height, maximum weight limit, and type (drawbridge, suspension, etc.). Examples of procedures for a bridge class could include opening and closing.

The packaging of an object's data with its procedures is known as *encapsulation*. Encapsulation allows modifications/additions to the system with minimal impact on other system components. This property is crucial for the successful development and maintenance of complex software systems such as GIS. Other major concepts for understanding OO models include *inheritance* and *polymorphism*. Inheritance is the automatic inclusion of attributes and methods from classes defined at a higher level in a class hierarchy. The class hierarchy is designed with more general classes being placed at higher levels and more specific classes being placed at lower levels. Inheritance reduces the need to duplicate data and code, while the structuring of a class hierarchy provides a logical organization of object "types."

Polymorphism allows methods for different objects to have the same name, while providing different implementations. For example, both a circle and a square class could implement a method that calculates area. Both methods could be named area, but the implementations would use the appropriate formula for either a circle or a square. Methods that invoked the area method for an object would use one name--the class of the receiving object would determine which implementation to use. Polymorphism helps to simplify system design, and reduces coding complexity by eliminating error-prone if-then-else type-checking constructs.

These concepts are applied to our work in the following ways. Encapsulation allows each geographic feature to maintain its own state of knowledge related to information and procedures necessary for conflation with another feature. Changes in the algorithms/implementations of a particular feature's conflation abilities do not affect other features. Inheritance allows us to describe general types of conflation knowledge applicable to multiple feature classes within a single class at a high level in the hierarchy. Lower-level classes that are more specific then automatically inherit this knowledge. Finally, polymorphism allows us to implement general conflation procedures that are valid for instances of any feature class; thus, polymorphism greatly reduces the need to be concerned with the issue of "type." The result is that we are able to treat, for example, railroad features and building features in the same manner at a logical level. Figure 2 shows a simplified class hierarchy for conflation.

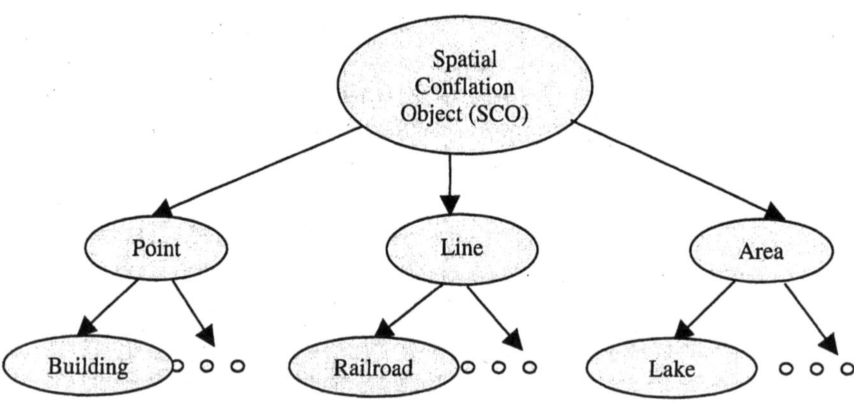

Figure 2. General class hierarchy for conflation.

The Naval Research Laboratory's Geospatial Information Database (GIDB) prototype [64], within which proof-of-concept implementation of this model is currently being performed, is centered on the concept of spatial range queries, also known as area-of-interest (AOI) queries. Given an AOI, through definition of manual or graphical bounding box coordinates or geographical place name, the GIDB is able to return any vector, raster or multimedia data available for that area. Advanced queries, such as those limiting vector data attribute values, are also available.

Our conceptual model of the conflation process is based on this system model of AOI queries, limited to the consideration of vector data only. A diagram of the conflation process is shown in figure 3. The primary point to note is that the process takes place at the individual feature level, irrespective of that feature's physical residence, or logical database, library or coverage inclusion. In the first step, the user selects an AOI. The query manager then retrieves all feature objects from the distributed database that fall within the AOI (subject to any other constraints imposed by the user). From this collection of objects, the query manager randomly selects one object to send a "conflate" message. That object follows the protocol for determining which, if any, of the remaining objects in the query collection are matching representations for itself. Any object determined to be a match is placed in the matching feature set for the conflating object, ranked according to similarity scores. Objects that have been placed in a matching feature set are removed from the general query collection, so as not to be candidates for subsequent conflation iterations. This philosophy implies that only those objects that have scores *strongly* suggesting matching features are placed in the matching feature sets. The query manager continues by sending "conflate" messages to the remaining members of the query collection. When the process is complete, the query manager returns the query collection for display. For any features with non-empty matching feature sets, the member of the set with the highest quality score (NOT the same as the similarity score) is returned as the representative for that feature. The quality score is an indication of the fitness for use of the data, and is based on multiple factors derived from both the data and metadata.

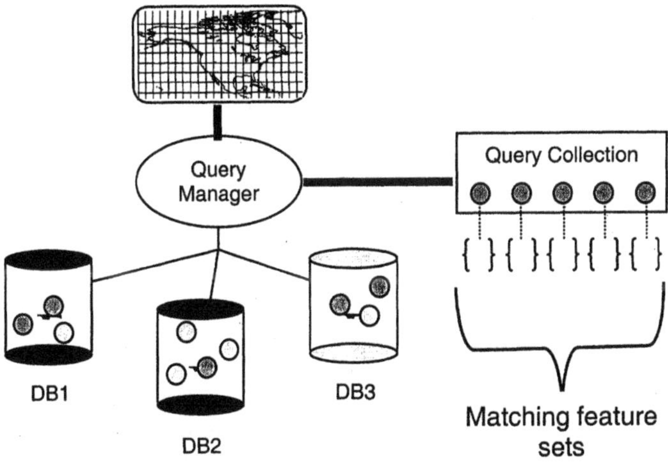

Figure 3. Conceptual model for distributed conflation.

Several points on the conceptual model are worth noting at this point. First, the responsibility for conflation is distributed. Each individual feature contains the inherited knowledge needed for performing feature matching for that particular class of geographic objects. This is preferable to a centralized conflation system, where a single conflation object "manages" the process. Such a system would be more difficult both to implement and maintain due to differences in the ways in which objects from various feature classes should be matched, as well as the ramifications of adding new features and/or system nodes for a distributed system. Second, the process is automatic. Because the query manager collects the features satisfying the user query and invokes the conflation method for each object before returning the final set, the user does not have to explicitly request conflation, or even need to know that conflation is an issue. Third, when the results are returned to the user, only a single representation of each object is presented. For now, our treatment of deconfliction is simply to select the "best" feature from a matching feature set, based on various objective and subjective criteria. This idea of simplifying the user's concern and involvement with conflation is critical for end-user based systems, as the whole need for such an automated approach is derived from the concept of non-expert users. A more detailed explanation of the model can be found in [63].

Though irrelevant at a conceptual level, physical allocation of data in a distributed system is of paramount concern for performance issues. Here, we briefly consider one possible allocation scheme and the potential impact as related to the model.

The scheme we consider is a partitioning based on representational classes. That is, all line features such as railroad lines, power lines, etc., reside on the same physical node; all point features are likewise allocated to the same node, as are all area features. In consideration of the hierarchy given in figure 2 within this setup, we believe the optimal partitioning of the hierarchy is to include all level 3 (feature class) classes and their instantiations together with their corresponding level 2 superclass. For example, all transportation lines, utility lines, etc. would be co-resident with the line class object. For conflation, this setup would minimize network traffic needed for transfer of data and execution of methods, as most of the specific conflation knowledge is inherited from the SCO and values instantiated at these lower two levels. Of course, other partitionings may also provide acceptable performance, but this one is given as an example of considerations between data allocation methods and the distributed conflation model.

5 Concluding Remarks

The importance of uncertainty modeling for spatial data and GIS is well recognized in the geographic information science research community. We have surveyed the relevant issues and some specific efforts at spatial modeling using fuzzy set approaches. As distributed systems and web-based interoperability have come to the forefront, uncertainty issues are then reflected in such an environment.

Our current research directions are to more fully develop formal approaches to distributed spatial data semantics and to produce prototypes based on these approaches.

Acknowledgments

This work was sponsored by the Marine Corps Warfighting Lab (MCWL) under Program Element 63640M, with LtCol David Durham, LtCol Jerry Brown, and MAJ Mike Berigan serving as program managers.

References

1. E. Codd, "A Relational Model for Large Shared Data Banks," *Communications of the ACM*, Vol.13, pp.377-387, 1070.

2. B. Buckles and F. Petry, "A Fuzzy Model for Relational Databases," *Int. Jour. Fuzzy Sets and Systems*, Vol. 7, pp. 213-226, 1982.
3. S. Shenoi and A. Melton, "Proximity Relations in Fuzzy Relational Databases," *Int. Jour. Fuzzy Sets and Systems*, Vol. 31, pp. 287-296, 1989.
4. M. Anvari and G. Rose, "Fuzzy Relational Databases," *The Analysis of Fuzzy Information Vol. II*, (ed. J. Bezdek), pp. 203-212, CRC Press, Boca Raton, FL, 1987.
5. M. Umano, "FREEDOM-0 : A Fuzzy Database System," *Fuzzy Information and Decision Processes*, (eds. M. Gupta and E. Sanchez), North-Holland, Amsterdam, pp. 339-347, 1982.
6. H. Prade and C. Testemale, "Generalizing Database Relational Algebra for the Treatment of Incomplete/Uncertain Information and Vague Queries," *Information Sciences*, Vol. 34, pp. 115-143, 1984.
7. M. Zemankova and A. Kandel, "Implementing Imprecision in Information Systems," *Information Sciences*, Vol. 37, pp. 107-141, 1985.
8. E. Rundensteiner, L. Hawkes, and W. Bandler, "On Nearness Measures in Fuzzy Relational Data Models," *Int. Jour. Approximate Reasoning*, Vol. 3, pp. 267- 298, 1989.
9. J. Medina, O. Pons, and M. Vila, "Gefred: A Generalized Model to Implement Fuzzy Relational Databases," *Information Sciences*, Vol. 47, pp. 234-254, 1994.
10. R. de Caluwe, ed., *Fuzzy and Uncertain Object-Oriented Databases*, World Scientific, Singapore, 1997.
11. M. Goodchild and S. Gopal, eds. *The Accuracy of Spatial Databases*, Taylor and Francis, Basingstoke, UK, 1990.
12. __, *ARC /INFO User's Guide: ARC/INFO 6.0 Data Model: Concepts and Key Terms*, Environmental Systems Research Institute, Redlands, CA.
13. D. MaGuire, "An Overview and Definition of GIS," *Geographical Information Systems: Principles and Applications, VOL 1 - Principles*, (eds. D. MaGuire, M. Goodchild, and D. Rhind), pp. 9-20, Longman, Essex GB, 1991.
14. University Consortium on Geographic Information Science, "Research Priorities for Geographic Information Science," *Cartography and Geographic Information Systems*, Vol. 23, No. 3, pp 115-127, 1996.
15. H. Veregin, "A Taxonomy of Error in Spatial Databases," Technical Report 89-12, National Center for Geographic Information and Analysis, Santa Barbara, CA, 1989.
16. P.A. Burrough, "Fuzzy Mathematical Models for Soil Survey and Land Evaluation," *Journal of Soil Science*, 40:477-492, 1989.
17. M.F. Goodchild, G. Sun, and S. Yang. "Development and Test of an Error Model for Categorical Data," *International Journal of Geographical Information Systems*, 6(2):87-104, 1992.
18. P.F. Fisher, "First Experiments in Viewshed Uncertainty: Simulating the Fuzzy Viewshed," *Photogrammetric Engineering and Remote Sensing*, 58:345-352, 1992.
19. P.A. Burrough and A.U. Frank, editors, *Geogruphic Objects with Indeterminate Boundaries*, Taylor and Francis, London, 1996.

20. P.F. Fisher, "Probable and Fuzzy Models of the Viewshed Operation," in M.F. Worboys, editor, *Innovations in GIS 1,*Taylor and Francis, London, pp. 161-175, 1994.
21. P.F. Fisher, "Boolean and Fuzzy Regions," in P.A. Burrough and A.U. Frank, editors, *Geographic Objects with Indeterminate Boundaries,* Taylor and Francis, London, pp. 87-94, 1996.
22. S. Kennedy, "The Small Number Problem and the Accuracy of Spatial Databases," Chapter 16, *The Accuracy of Spatial Databases*, (eds. M. Goodchild and S. Gopal), Taylor and Francis, Basingstoke, UK, 1990.
23. D. Stoms, "Reasoning with Uncertainty in Intelligent Geographic Information Systems," *Proc. GIS 87 - 2nd Annual Int. Conf on Geographic Information Systems*, 693-699, American Soc. for Photogrammetry and Remote Sensing, Falls Church VA, 1987.
24. M.F. Goodchild, D.R. Montello, Peter Fohl, and Jon Gottsegen, "Fuzzy Spatial Queries in Digital Spatial Data Libraries," *Proc. FUZZ-IEEE '98*, Anchorage, AK, pp. 205-210, 1998.
25. C.R. Ehlschlaeger, A.M. Shortridge, and M.F. Goodchild, "Visualizing Spatial Data Uncertainty Using Animation," *Computers and Geosciences,* 23(4):387-395, 1997.
26. V. Robinson and A. Frank, "About Different Kinds of Uncertainty in Geographic Information Systems," *Proc. AUTOCARTO 7 Conference*, 1985.
27. V. Robinson, "Implications of Fuzzy Set Theory for Geographic Databases," *Computers, Environment, and Urban Systems*, Vol. 12, pp. 89-98, 1988.
28. V. Robinson, "Interactive Machine Acquisition of a Fuzzy Spatial Relation," *Computers and Geosciences*, Vol. 6, pp. 857-872, 1990.
29. C. Giardina, "Fuzzy Databases and Fuzzy Relational Associative Processors," Technical Report, Stevens Institute of Technology, Hoboken NJ, 1979.
30. J. Baldwin, "Knowledge Engineering Using a Fuzzy Relational Inference Language," *Proc IFAC Symp. on Fuzzy Information Knowledge Representation and Decision Analysis,* pp. 15-21, 1983.
31. L. Zadeh,"Test-Score Semantics for Natural Languages and Meaning Representation via PRUF," *Empirical Semantics*, (ed. B. Rieger), Brockmeyer, Bochum, GR, pp. 281-349, 1981.
32. V. Cross, "Using the Fuzzy Object Model for Geographical Information Systems," *Proc. 18th Intl. Conference of the North American Fuzzy Information Processing Society(NAFIPS)*, pp. 814-818, June 10-12, 1999.
33. E. L. Usery, "A Conceptual Framework and Fuzzy Set Implementation for Geographic Features," in *Geographic Objects with Indeterminate Boundaries*, ed. P. Burrough and A. Frank, Taylor and Francis, London, pp. 71-86, 1996.
34. Y. Leung, *Spatial Analysis and Planning Under Imprecision*, Amsterdam, Elsevier, 1988.
35. M.H. Katinsky, *Fuzzy Set Modeling in Geographic Information Systems*. Unpublished Master's Thesis, Department of Geography, University of Wisconsin-Madison, Madison, WI, 1994.

36. T. Sarjakoski, “How Many Lakes, Islands, and Rivers are there in Finland: a Case Study of Fuzziness in Extent and Identity of Geographic Objects,” *Geographic Objects with Indeterminate Boundaries*, ed. P. Burrough and A. Frank, Taylor and Francis, London, pp. 299-312, 1996.
37. A.U. Frank and M.F. Goodchild, "Two Perspectives on Geographic Data Modelling," Technical Paper 90-11, National Center for Geographic Information and Analysis (NCGIA), 1990.
38. H. Couclelis, "Beyond the Raster-Vector Debate in GIS," in Frank, A.U., Campari, I. And Formentini, U. (Eds), *Theories and Methods of Spatio-Temporal Reasoning in Geographic Space*, Lecture Notes in Computer Science 639, Berlin: Springer, pp.65-77, 1992.
39. V. B. Robinson, "A Perspective on Managing Uncertainty in Geographic Information Systems with Fuzzy Sets," *Proc. FUZZ-IEEE '98*, pp. 211-215, 1998.
40. T.J. Davis, and C. P. Keller, "Modelling Uncertainty in Natural Resource Analysis Using Fuzzy Sets and Monte Carlo Simulation: Slope Stability Prediction," *International Journal of Gegraphical Information Systems*, 11(5): 409-434, 1997.
41. D. Altman, "Fuzzy Set Theoretic Approaches for Handling Imprecision in Spatial Analysis," *International Journal of Geographical Information Systems*, 8(3): 409-434, 1994.
42. M. Cobb, *An Approach for the Definition, Representation and Querying of Binary Topological and Directional Relationships between Two-Dimensional Objects*, Ph.D. dissertation, Tulane University, New Orleans, LA, 1995.
43. M. Cobb. and F. Petry, "Modeling Spatial Relationships within a Fuzzy Framework," *Journal of the American Society for Information Science*, 49(3), pp.253-266, 1998.
44. M. Cobb. and F. Petry, "Fuzzy Querying of Binary Relationships in Spatial Databases," *Proc. IEEE Intl. Conference on Systems, Man and Cybernetics*, Vancouver, BC, pp. 3624-3629, 1995.
45. M.J. Egenhofer, and R. Franzosa, "Point-Set Topological Spatial Relations," *Intl. J. Geo. Info. Sys.*, Vol. 5, No. 2, pp. 161-174, 1991.
46. M. Cobb, and F. Petry, "Integration of a Fuzzy Query Framework with Existing Spatial Query Languages," *Proc. Fifth IEEE Intl. Conf. On Fuzzy Systems* (FUZZ-IEEE), New Orleans, LA, pp. 93-99, 1996.
47. M. Nabil, J. Shepherd and A.H.H. Ngu, "2D Projection Interval Relationships: A Symbolic Representation of Spatial Relationships," *Advances in Spatial Databases: 42nd Symposium*, SSD '95, pp. 292-309, 1995.
48. J. Sharma, and D.M. Flewelling, "Inferences from Combined Knowledge about Topology and Direction," *Advances in Spatial Databases: 42nd Symposium*, SSD '95, pp. 279-291, 1995.
49. E. Clementini, J. Sharma and M.J. Egenhofer, "Modelling Topological and Spatial Relations: Strategies for Query Processing," *Computers and Graphics*, Vol. 18, No. 6, pp. 815-22, 1994.

50. H.-P. Kriegel, M. Schiwietz, R. Schneider, and B. Seeger. "Performance Comparison of Point and Spatial Access Methods," in A. Buchmann, O. Gunther, T. Smith & T. Wang (Eds.), *Design and Implementation of Large Spatial Databases*, Lecture Notes in Computer Science 409, Santa Barbara, CA: Springer-Verlag, pp. 89-114, 1989.
51. H. Samet, *The Design and Analysis of Spatial Data Structures*, Reading MA: Addison-Wesley, 1989.
52. S.K. Chang, et al. "An Intelligent Image Database System," *IEEE Transactions on Software Engineering*, Vol. 14, No. 5, pp. 681-688, 1988.
53. J. F. Allen, "Maintaining Knowledge about Temporal Intervals," *Communications of the ACM*, Vol. 26, No. 11., pp. 832-843, November 1983.
54. A. Saalfeld, "Conflation: Automated Map Compilation," *International Journal of GIS*, 2(3):217-228, 1988.
55. M. Godau, "A Natural Metric for Curves--Computing for Polygonal Chains and Approximation Algorithms," *Proc. Symposium on Theoretical Aspects of Computer Science*, STACS '91, Springer Lecture Notes in Computer Science, V. 480, pp. 127-136, 1991.
56. D. Siegel, "A non-traditional solution to conflation," *Proceedings of the Urban and Regional Information Systems Association annual conference,* Washington, D.C., 16-20 July, pp. 462-75, 1995.
57. L. A Zadeh, "Fuzzy Sets," *Information and Control*, pp. 338-353, 1965.
58. G. Shafer, *A Mathematical Theory of Evidence*, Princeton University Press, 1976.
59. M. A. Cobb, M.J. Chung, V. Miller, H. Foley III, F. E. Petry, K. B. Shaw, "A Rule-Based Approach For The Conflation of Attributed Vector Data," *GeoInformatica*, 2(1), pp. 7-35, 1998.
60. H. Foley, F. Petry, M. Cobb and K. Shaw, "Utilization of an Expert System for the Analysis of Semantic Characteristics for Improved Conflation in Geographic Information Systems," *Proc. of the 10th Intl. Conf. On Industrial and Engineering Applications of AI*, Atlanta, GA, pp. 267-275, 1997.
61. H. Foley, F. Petry, M. Cobb and K. Shaw, "Using Semantic Constraints for Improved Conflation in Spatial Databases," in *Proc. 7th Intl. Fuzzy Systems Association World Congress*, Prague, 193-197, 1997.
62. E. Lim, J. Srivastava and S. Shekar, "An Evidential Reasoning Approach to Attribute Value Conflict Resolution in Database Integration," *IEEE Transactions on Knowledge and Data Engineering*, 8(5):707-723, 1996.
63. M. Cobb, F. Petry and K. Shaw, "Uncertainty Issues of Conflation in a Distributed Environment," *Proceedings of GIS/LIS '98*, Fort Worth, TX, pp.436-448,Nov.10-12,1998.
64. M. Cobb, H. Foley III, R. Wilson, M. Chung and K. Shaw, "An OO Database Migrates to the Web," IEEE Software, 15(3), pp. 22-30, May-June.

On Fuzzy Sets and the Management of Uncertainty in an Intelligent Geographic Information System

Vincent B. Robinson [1]

[1] Department of Geography, University of Toronto, Mississauga, ONT L5L 1C6. Canada

Abstract. The Knowledge Based Land Information Manager and Simulator (KBLIMS) is used as the basis for providing a perspective on using fuzzy sets as the basis for managing certain kinds of uncertainty in a geographic information system. A brief perspective is presented on the implications for the classification of landscape data, object model, and query model of the system. Portions of the system may exploit the rich research of fuzzy sets to address very specific problems such as model self-evaluation and control. Addressing the issue of representation and management of uncertainty at both the landscape analysis and query level leads to a process of formalizing and managing uncertainty that cascades through the system warranting changes at the object model level subsequently forcing substantial additions to the knowledge base.

Keywords. Geographic information systems, uncertainty, object modeling.

1 Introduction

Geographic Information Systems (GIS) may offer the most significant opportunity for real-world application of fuzzy sets and fuzzy logic [1, 2]. Fuzzy sets have been used to address a wide variety of problems within Geographic Information Systems. Since [3] there has been a substantial increase in the use of fuzzy sets in the field of geographical information science. Usually the use of fuzzy sets is reported in journal or proceedings papers within the context of addressing a specific problem in geographic information processing or representation. This paper attempts to provide a higher level perspective. The approach taken here is to begin with a knowledge based geographic information system, which serves as a computational laboratory and at this time, does not incorporate fuzzy sets. The system provides a framework about which various aspects of managing uncertainty in a complex geographic information processing environment may be discussed. Although there are many kinds of uncertainty that are relevant in the context of intelligent geographic information systems, this paper will consider those that are appropriately considered amenable to management using fuzzy sets.

In the Knowledge based Land Information Manager and Simulator (KBLIMS) information is organized and accessible with reference to distinct, identifiable landscape units [4]. Object-oriented and knowledge based techniques are exploited to provide suitable tools for simulating ecosystem processes on the basis of high order geographic features such as hillslopes, streams, and forest stands. Ecological simulation processes are organized around a watershed-based model of terrain with an infinite hierarchy of watersheds, hillslopes and stream links. Spatial representation at each level is organized into a network, consisting of a set of stream links, stream-to-hillslope adjacency links, and hillslope-to-hillslope adjacency links (Figure 1). This graph structure is the organizing principle for its object-oriented spatial database [5]. KBLIMS serves as a useful geographic information science laboratory because it can be used to address questions of representation, analysis, physical process modeling, and intelligent query within a rich problem domain. This problem domain is relevant to a broad field such as forest ecosystem modeling which demands the management of geographically-based information detailing complex interactions between climatic, topographic, hydrologic, pedological and ecological processes.

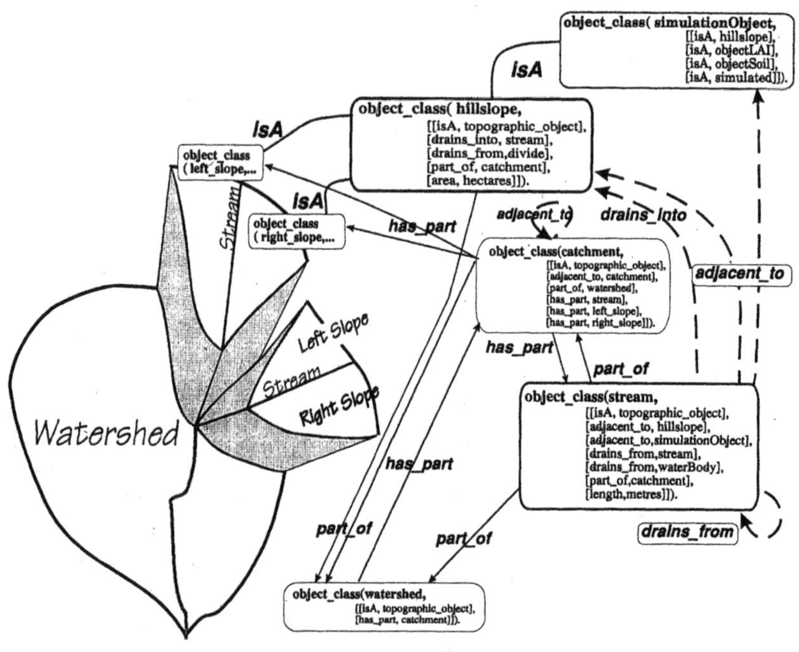

Figure 1. Representation of a watershed in KBLIMS.

2 Classifications and Landscape Heterogeneity

Terrain objects are used to define the landscape units upon which KBLIMS is organized. To arrive at these terrain objects it is necessary to conduct terrain analysis of image and map data. In KBLIMS the system translates layer-based geographic information into object-based geographic information. The terrain partitioning approach taken in KBLIMS can adaptively scale terrain partitioning from a few, large hillslopes to many, small hillslopes for any given drainage basin. It is the scale flexibility that allows a layer-to-object transformation without significant loss of land surface heterogeneity meaningful to ecosystem simulation [4,6]. These ecosystem process models are made up of many submodels.

The attributes of the landscape units are of particular importance to the scientist-user since the instances of attributes are necessary input data for the ecosystem models. These attributes are often the result of "natural resource mapping" efforts. Results of such efforts are typically described as soil maps, or vegetation maps, *etc.* Each spatial unit, raster cell or vector polygon, is usually classified according to how well it fits into particular crisp set such as soil type. These classifications are used to describe how the properties of the land surface vary over space. Thus, the classes themselves are often surrogates for other properties such as the hydraulic properties of a soil type. However, it has become clear that these classes are often not crisp in attribute space or geographic space. Hence, the use of fuzzy sets to address the problem of uncertainty in natural resource mapping (*e.g.* [7, 8])

2.1 Soils, Vegetation, and Fuzzy Representation

In ecosystem models like the one used in KBLIMS soils information, especially its related hydraulic properties, is critical to driving several submodels. The spatial distribution of soils, and their associated properties, is a necessity. Given its importance it is not surprising the amount of research that has concentrated on the uncertainty related to soils maps [7, 9,10]. Therefore the soils input to an ecosystem model would be some representation of the spatial distribution of soils, particularly its associated properties. Approaches that exploit fuzzy sets, GIS, remote sensing, and artificial intelligence techniques in combination with the spatial distribution of environmental variables to arrive at a fuzzy representation of the spatial distribution of soils have improved on the more traditional methods [7,9]. Figure 2 illustrates the major components in the process used in [9] to arrive at soil similarity vectors that have a spatial distribution. Like MacMillan *et al* [7], Zhu [9,10] uses soil experts to develop a base of knowledge on the relationships between soil series and the environmental variables elevation, slope gradient, slope aspect, profile curvature, canopy coverage, and parent material. This knowledge base is used in conjunction with environmental data that varies spatially over a particular region. The combination of the knowledge base with the spatially distributed environmental data is done using a fuzzy inference engine. For each location in a raster database, the fuzzy inference engine takes the set of environmental conditions for the location and combines them with the relationship functions for a soil series to compute the similarity value of the local soil to the more general soil series (see Figure 2). The result of the integration is a set of soil similarity vectors distributed over a geographic area.

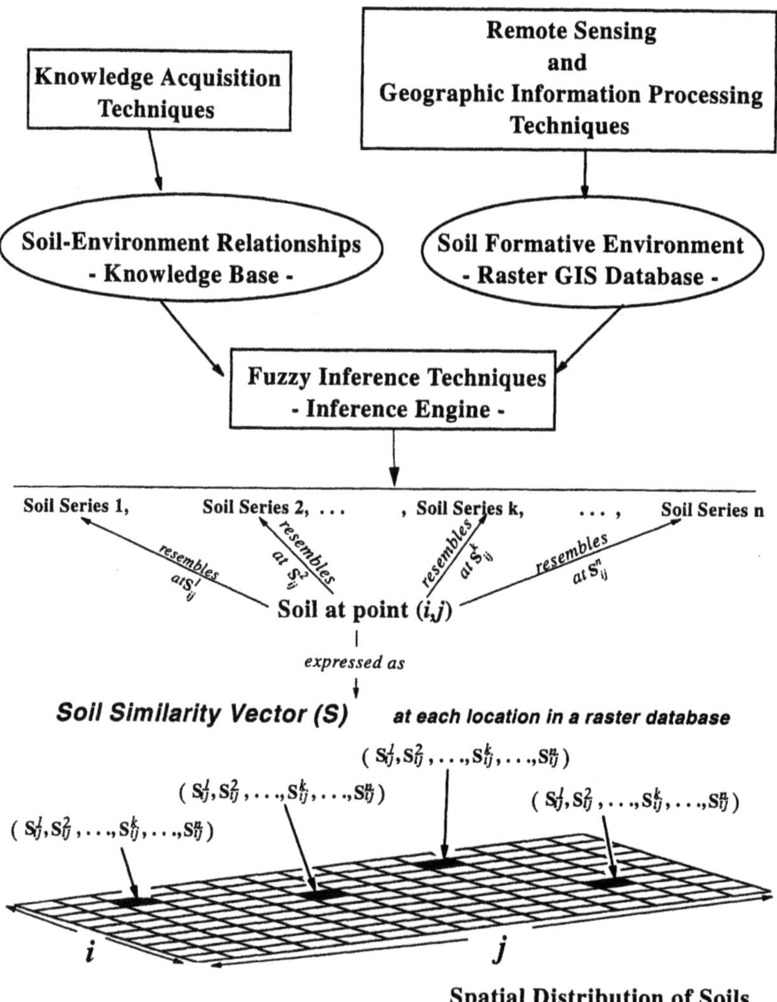

Figure 2. Knowledge based process of deriving soil similarity vectors that are used to arrive at a fuzzy representation of soil information (adapted from [9]).

The fuzzy knowledge based approaches (*e.g.* [7,9]) clearly provide more accurate representations of how soils vary over space, hence their relevant properties as well. The soil similarity vectors are "hardened," *i.e.* defuzzified, to produce a soil map. For example, a comparison between the soil series observed in the field and two sets of soil series from soil maps reveals the results of this fuzzy knowledgebased approach improved accuracy by nearly 20%. A similar magnitude of improvement was observed when the soil A-horizon depth was inferred from the soil similarity vectors [9]. It should be noted that the evaluation of "more accurate" requires the conversion from a fuzzy representation of local soils to a crisp representation. The defuzzification of these fuzzy representations is also mandated by the representational contraints imposed by existing geographic information systems and/or the process models that would use the soil information.

In a similar manner, vegetation characteristics, such as leaf area index (LAI), are important in some portions of the hydroecological simulation system that is a part of KBLIMS. Fuzzy modeling of vegetation from remotely sensed imagery typically concentrates on obtaining a fuzzy representation of a land cover map (*e.g.* [7,8]). Like the construction of soil maps, these approaches have generally been ends in themselves. That is to say that the objective has been to basically arrive at a map of land cover classifications. Generally speaking such efforts have rarely been incorporated into anything more than a simple retrieval system. Thus, it is common for the fuzzy land cover classifications to be defuzzified before being utilized in either a GIS or spatially explicit model.

2.2 Fuzzy Relationships and Conflation

The cases of soil and land cover classification are basically cases of developing data to populate a geographic database. However, there are many situations when there is a geographic database already in existence. Its spatial coverage may be partial or the certainty of the data may not be uniformly high over its spatial extent. Given the existence of a geographic database(s) covering the same area, conflation may be needed to update the geographic database without losing legacy data. Typically, conflation is regarded as the combination of information from two digital maps to produce a third that is superior to either of its component sources. Conflation can also be used when one map source may be more accurate with respect to positional or attribute information or one map source contains information, such as geographic features, missing in another.

Recently a process using extensions to the minimum bounding rectangle (MBR) model to aid in the reliable determination of fuzzy spatial relationships has been applied to the case of conflation. The ability to reliably handle fuzzy spatial relationships provides an improved method of conflation [11]. This translates into a geographic database in which there is higher level of confidence. In other words, the resulting geographic database will be a superior representation of the region and its attributes than before conflation.

2.3 Fuzzy Representation and Model Integration

In an intelligent geographic information system, such as KBLIMS, there is a close relationship between representation and ecosystem process models. In such systems there is a need to be able to incorporate the uncertainty represented in the above approaches in such a manner that it is of some use to the hydroecological simulation system as well as the intelligent query portion of the system.

In this sense, the uncertainty model proposed by Davis and Keller [12] illustrates how one may incorporate fuzzy metadata concepts with Monte Carlo simulation to eventually support a system that can use the uncertainty information to detail how much confidence to place in the model output.

Although the above approaches are generally couched in terms of raster-based data, a strategy has been presented for representing fuzzy geographic boundaries in a vector-based data model. Both approaches are in general layer-based and their role is substantially the same [13]. They arrive at a fuzzy representation of what has traditionally been represented in a crisp manner. The significance in terms of an intelligent GIS such as KBLIMS is that the information is ultimately used to populate instances of objects. Therefore, use of these fuzzy approaches begs the question of how, or if, the additional granularity afforded by the fuzzy representation can be exploited in a meaningful manner through the object model underpinning KBLIMS.

3 Object Modeling

Object modeling in an intelligent geographic information system provides the ability to encapsulate a geographical object as a unit of spatial and attribute information along with its associated methods [*e.g.* 4,7,14].

3.1 Object Model of KBLIMS

In KBLIMS the class schema is defined as **object_class** = $<\Phi, \Gamma, \Sigma>$ where Φ is the class name uniquely identifying an object class, Γ is a finite set of inheritance classes defined by the relation ***isA*** on Φ, and Σ is a finite set of attribute descriptions on Φ. Each attribute description, $(\rho,\phi) \in \Sigma$, consisting of a predicate, ρ, and a referent class, ϕ, is used to automatically generate a query on the knowledgebase.

Aggregation is used to express relationships between objects and describe attributes of an object using the form:

object_class(object,[[attribute,value]]).

This triplet declares that an individual (object) has a relationship (attribute) to some value or other individual (value). For example, The definition of the object_class describing topographic objects is

object_class(topographic_object,
[[elevation,meters],
[aspect,degrees],
[gradient,degrees]])

Figure 1 illustrates how the object_class triplet is used in conjuction with ***isA*** relations to represent topographic objects within KBLIMS [4].

The basis of the object-oriented paradigm is that crisp sets of objects can be used to model real-world problems [14]. It can be clearly shown that KBLIMS does follow that paradigm (*e.g.*, see Figure 1). Hillslopes, catchments, wetlands and other objects relevant to ecosystem modeling are represented and manipulated as members of crisp sets. At this juncture then the question becomes how can fuzziness at the object schema level be accommodated. Hadzilacos and Tryfona [15] suggest that fuzziness at the schema level can be accommodated by the addition of geometry type fuzzy_point. However, this *ad hoc* extension is tantamount to defining object classes as subclasses of ordinary classes. The approach may work to address relatively small problems of fuzziness in geographic databases, but it avoids the multitude of issues raised when considering the incorporation of fuzzy objects at all levels of a geographic information system. In contrast, a recent effort has considered extending ERD modeling notation to develop a GIS data model of fuzzy relations among GIS datasets [16]. Thus, there are several researchers looking at the problem of fuzziness in GIS as a fundamental problem that is addressed only by its incorporation, in a pervasive manner, throughout the object model [1,2,14,16]

3.2 Representing Fuzzy Objects

The implication of using methods like those used in [7,8,9,17] to populate geographic databases is that there will always be uncertainty regarding the categorization of instances. Rather than defuzzifying the data to fit the usual object model paradigm, the alternative is to represent the fuzziness explicitly. In a system such as KBLIMS, which enforces the usual object model paradigm, the question may become how to weaken such enforcement to accommodate uncertain instance-of relations. In addition, there may well be a need to allow for uncertain *isA* relations.
One approach might be to introduce a fuzzy-instance-of class to address the uncertain relationships between classes and instance objects; a fuzzy-relationship class for managing the uncertain relationships among instance objects; and a fuzzy-class-hierarchy for managing the uncertainty of the class hierarchy (*e.g.* uncertain *isA* relations) [18].

3.2.1 Representing Fuzzy Attributes of Objects

Using examples based on the Object Data Management Group (ODMG) object definition language (ODL) the object model used in KBLIMS can be contrasted with the fuzzy extensions to ODL discussed in [14]. One extension would permit the domain of an attribute type in an object type definition to specify a fuzzy set collection of literals, user-defined object types, or other collection types [14]. In fuzzy extensions to the ODMG object model, an uncertain value for single valued property is allowed. For example, the attribute aspect for a topographic object might be known as an inexact value such as *south-facing*. Cross and Firat [14] propose a fuzzy extension to the ODMG object data model that would specify that the value of a single valued property of an object contains uncertainty, *i.e.,* the collection type Fuzzy_value. For example, if uncertainty is present in values for the attribute gradient, then the attribute type could be specified to hold a fuzzy value such as Fuzzy_value<degrees>.

```
interface topographic _object {
        extent topographic_objects;
        attribute location relative_position;
        attribute meters elevation;
        attribute Fuzzy_value<degrees> aspect;
        attribute Fuzzy_value<degrees> gradient;
        relationship Fuzzy_set<region_unit> is_part_of
        inverse region_unit:: has_topology;
        add_to_region(in region_unit);
remove_from_region(in region_unit);
```

Figure 3. Example of fuzzy extension of ODMG object model (adapted from [14]).

3.2.2 Representing Fuzzy Relations of Objects

Although useful, this simple extension is inadequate for all possibile situations in an intelligent geographic information system such as KBLIMS. To illustrate this point [14] suggest considering that the relationship *is_part_of* could be specified as a fuzzy set value where the membership degree determines to what degree the topographic object lies within the boundaries of the region. The definition becomes *Fuzzy_set<region_unit> is_part_of.*

If revisions to the ODMG object model permit properties of objects to become objects themselves, then an object representing a fuzzy linguistic variable could be created. It has been pointed out that such an object would have its own properties, *e.g.*, its own universe of discourse and linguistic term set. Permitting the type of an object property to be a fuzzy linguistic variable would specify that the only values for this property would be from the term (see [14]). Since linguistic terms are one of the more common ways of utilizing fuzzy sets in geographic information systems (*e.g.* see [19,20]), this revision would helpful in standardizing the implementation of linguistic variables in a fuzzy geographic information system.

In KBLIMS ***isA*** relations are used extensively to represent and manage the subtype/supertype hierarchy. For example, waterBody represents a generalization of several different kinds of water bodies (Figure 4).

```
object_class(waterBody,
        [[adjacent_to,hillslope],
        [drains_from,stream],
        [drains_into,stream],
        [surface_area,hectares],
        [elevation,meters]]).

object_class(wetland,
        [[isA,waterBody]]).

object_class(lake,
        [[isA,waterBody]]).
```

Figure 4. Using *isA* relations in KBLIMS object model to represent supertype/subtype relations

To illustrate the use of ***isA*** hierarchies to manage semantic heterogeneity, Robinson and Mackay [4] present example of wetlands and lakes. In keeping with the crisp set basis of the object model paradigm the wetland and lake subtypes are crisp subtypes of supertype waterBody. However, wetland is an area that is not strictly water and not strictly land. In other words, the degree to which it is a waterBody can be thought of as something other than crisp. It has been suggested how this might be represented using extensions to the ODMG object model [14] (Figure 5).

```
interface waterBody:topographic _object {
        extent water_bodies;
        attribute Fuzzy_value<hectares> surface_area;
        attribute meters maximum_depth;
        attribute meters minimum_depth;
        relationship Set<stream> drains_into
        inverse stream:: drains_from;
        relationship Set<stream> drains_from
        inverse stream:: drains_into;
}
interface wetland:waterBody with 0.6 membership {
        extent wetlands;        // Name of extent
}
interface lake:waterBody {
        extent lake;            // Name of extent
}
```

Figure 5. Example of representing fuzzy supertype/subtype relations using extensions to the ODMG object model (adapted from [4]).

It is clear that the introduction of this degree of membership will affect the inheritance of attribute values, constraints, and methods. It is not clear what the rules governing this kind of inheritance should be. At this stage in our understanding of fuzzy object management the rules governing fuzzy inheritance may be context specific. In other words, the inheritance rules governing inheritance of attribute values, constraints, or methods from the supertype waterBody to the subtype wetland may be depend on the specific semantics of the problem domain.

3.3 Research Issues in Modeling Fuzzy Geographical Objects

A particularly thorny problem arises in part due to the nature of problem domains being represented using geographic information systems. It can be summarily described the problem of *between classes membership* vs *within class membership* Cross and Firat [14] illustrate this problem with an example drawn from an object base like that of KBLIMS. For each subtype of waterBody a fuzzy predicate over the attributes of *maximum_depth* and *minimum_depth* may be defined in terms of to what degree an object instance is considered a particular subtype of waterBody. This membership is different from the membership degree of the subtype wetland with respect to its supertype waterBody. Thus, membership of an object instance in waterBody could be affected both by the supertype/subtype membership degree between wetland/waterBody and by the object instance's membership within wetland.

Exactly how these uncertain instances and relations are handled is a major to be addressed explicitly by any intelligent geographic information system based on fuzzy extensions to the usual object model. As previously suggested, when dealing with fuzzy supertype/subtype relations, such as between waterBody and wetland, the inheritance rules governing inheritance of attribute values, constraints, or methods from the supertype waterBody to the subtype wetland may be depend on the specific semantics of the problem domain. These semantics will need to be explicitly represented and handled by the system. Since the approach taken in constructing KBLIMS is a query-based approach it is reasonable to suggest that many of the issues regarding the interpretation of fuzzy object models will be for the query model of KBLIMS to address. This also implies that a considerable amount of knowledge required to represent and manage the semantics may depend on the semantics of the problem domain.

4 Geographical Query Model

In KBLIMS a query is of the form: $\sigma_x = \Pi \times \rho(\Phi^b, \nu^b, \phi^b, \mathbf{X}^{f(\tau)})$ where Π is the knowledge base, and $\mathbf{X}^{f(\tau)}$ refers to a free variable which is instantiable over some type, τ, satisfying the predicate, ρ. Superscripts b and f represent bound and free arguments respectively. Since all spatial queries in KBLIMS are based on the watershed partitioning of the landscape ν is an integer value from a topological index (T) constrained such that $(\forall i \in \nu)((\exists j \in T) \rightarrow i = j)$ which describes the relationships among the stream links and hillslopes [5,6]. This query model has been used to develop a visual logic-based query manager [21].

4.1 Issues of Simple Geographical Query

In a system like KBLIMS fuzzy spatial relations may be relevant at a number of levels. Fuzzy spatial relations can be used to specify constraints on a query. Consider a simple example where the user must specify a crisp range of aspect values, as in Figure 6.

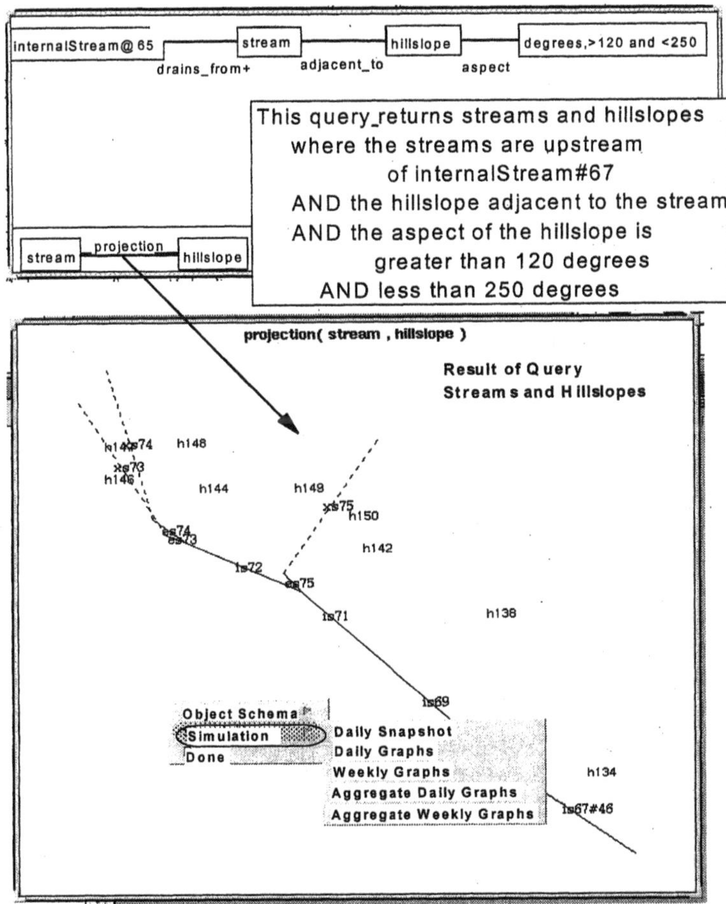

Figure 6. Example of complex geographical query in KBLIMS.

The fuzzy alternative would be for the user to be able to specify terms such as *south-facing*. The linguistic term would then be used to answer the query. It is worth noting that this example of a linguistic term constraint is consistent with example of using an object model, as in section 3.2.1, that allows for the representation of fuzzy attributes such as south-facing gradient. The alternative is having the query model implementation be capable of specifying the linguistic contraint and mapping it in relation to a set of crisp values describing a nonfuzzy attribute of an object.

The question usually rises at this point as to how to populate the fuzzy relation with membership values. One approach that has been used for spatial relations such as *near*, *far*, *etc.* is based on machine acquisition of a fuzzy concept through the use of nonfuzzy question/answer session with the user [22].

Although it has been suggested that the applications of GIS are facilitated if natural query languages can be used [19,23], it remains a questionable assertion in practical terms. However, there is much to be learned through research in the natural language query of geographical information that. Both the natural language query (NLQ) and the visual query (VQ) approaches typically create a query graph. The NLQ will typically defuzzify the terms as it maps from the linguistic terms, sentences, *etc.*, to a query language such as SQL [23] that is used to retrieve the information from a database management system. In contrast, VQ in KBLIMS maps to a logical expression that resembles a logic program. Thus, the VQ in KBLIMS can clearly be extended by incorporation of modification, composition, and quantification fuzzy formulae that have been used in NLQ with fuzzy sets [19,23]. The grammar would be simpler in the sense that it would be constrained by the logic of the semantic model.

If nothing else in KBLIMS were changed from its current, crisp object model state, the mapping would be from a fuzzified VQ to a set of logic programming statements rather than SQL. This mapping would require similar rules of defuzzification as for SQL. In addition, the mapping could be from a fuzzified VQ to a fuzzy SQL [24]. However, if the KBLIMS object model were extended to explicitly handle fuzzy attributes and relations, the mapping from the query to database would be more direct, more "natural," and would support more general types of uncertainty [25].

4.2 Retrieval of Geographic Information

Regardless of the model used to specify the terms in the query, there remains the necessity of having an automated method for retrieval of information from a geographic database. Even with a crisp, discrete representation of a geographic domain, as in KBLIMS, problems of determining relationships among various geographic objects may still arise. Such problems usually arise because of the difficulty automated systems have discriminating between similar spatial relationships and communicating subtle differences in those spatial relationships to the system (or user) [14]. It has been suggested that the use of fuzzy methods associated with linguistic variables is promising approach for duplicating human reasoning in the area of spatial relationship determination. [26,14].

One appraoch utilizes an abstract spatial graph (ASG) to represent a transformation of 2-dimensional space into 0-dimensional space [14]. KBLIMS has a query model is based on a similar concept. Since it is a visual query approach there are several points in the process where efficient spatial retrieval could become an issue. This is usually a significant issue when working with large complex landscapes. The minimum bounding rectangle (MBR) is a common approach when addressing retrieval issues through the logic of the spatial model. In a system such as KBLIMS the MBR approximation may lead to unsatisfying results if the query language is extended to include fuzzy terms. The fuzzy terms in the query language will need to translated to a form that is meaningful to the retrieval process. If MBRs are used but are maintained as the crisp approximations there may be difficulty in mapping from fuzzy spatial queries to crisp MBR approximations. ASG utilizes MBRs to represent features. It addresses some of the challenges posed in the modeling of topological spatial relations using MBR representations by not only representing topological and directional relationships, but also supplemental information needed for fuzzy query processing. For example, each node in an ASG has weights that can be used to define fuzzy qualifiers for the query languge [14]. Thus, an approach such as ASG could be used to extend a crisp spatial retrieval model such as in KBLIMS to one that explicitly incorporates fuzziness in the spatial graph.

4.3 Issues of Complex Geographical Query

The simple retrieval of geographical information (*i.e*,. combined spatial and attribute data) from KBLIMS can be accomplished with the same query model used for more complicated queries (Figure 6). These complex queries do not result in just the retrieval of information, they exploit a significant capability of an intelligent GIS such as KBLIMS which is its ability to automatically run complex ecosystem process simulations. In the construction of the visual query the system does not differentiate between those objects where simulation is needed to instantiate and those not based on simulation results. In [20], linguistic variables were used as the basis of a natural language interface that specified rules for a cellular automata model of land use development. This example illustrates how linguistic variables can used in a system interface to manage a spatially explicit simulation system [20]. The hydroecological simulations run by KBLIMS are based on an integration of several process models, hence there are a large number of parameters that are specified from field work or previous research. In terms of conducting sensitivity analysis on how results vary given varying inputs, a fuzzy linguistic variable approach would make it more useable, especially by novice users [22]. One of the implementation issues would be how to populate the linguistic variables with membership values and how will KBLIMS subsequently map between them and parameters specifying canopy dynamics, soil properties, *etc*. Although potentially an interesting exercise, this may in fact lead to more system overhead than can be justified. However, this raises an issue rarely discussed in the GIS or process modeling community, namely the automated control of complex models.

4.4 Ecosystem Models and Semantic Heterogeneity

Complex ecosystem process models typically have weak or poorly defined semantics. As a result, the output from these models is often difficult to interpret outside a narrow problem domain for which they were designed. Complex process model environments are characterized by a number of different models linked in some fashion to one another with each submodel describing some facet of the ecosystem. Therefore, determining semantic agreement between submodels is not just an information management issue but has implications that can affect scientific merits of the results [27].

In a query-oriented system, such as KBLIMS, the context for determining semantic agreement beween submodels may be provided by the query, **Q**:

$Q=<\alpha,\theta,\tau,\mu(\in_\alpha)>$ where α is the goal, θ and τ define the spatial and temporal domains, respectively, within which α is defined. The membership function $\mu(\in_\alpha)$ can describe the semantic proximity of α as determined by a collection of submodels. Proximity can be presented in linguistic terms understood by the model-user [28].

Terms such as sensitivity, predictability, and synchronicity, can portray the qualities of interest in a sensitivity analysis. Each term can be described by fuzzy memberships defined over a spatial domain and instantiated by parallel simulation of semantically erroneous and semantically corrected representations within the same integrated model. For example, a measureable semantic agreement in the water table position predicted by two spatially different submodels of RHESSysD can be computed as the sum, $\in_\phi$,of results of rules used for self-evaluation within RHESSysD. The physically-based constraint would be that water must flow from higher elevation points to lower elevation points within the bounds of a specific spatial domain. For instance, the patch-level water table position is adjusted to so that fine resolution and coarse resolution spatial models within RHESSysD are interoperable. However, hysteresis, capacitance, and other dynamic qualities within the model may prevent model interoperability given the user's context (α). For instance, the user context, *soil moisture*, may provide a map showing semantic distance between the interpretations provided by the submodels of RHESSysD. In this case, one could determine where the models give acceptably close interpretations using some criterion, or decision rule, to determine if two objects refer to the same thing. Thus, linguistic terms can present semantic proximity in a way that is meaningful to the end-user and reduces the cognitive demands on a user who is trying assemble environmental systems knowledge from existing disparate sources. [27,28]

This situation is one often overlooked in linking GIS with complex environmental modeling environments. However, it is this area where fuzzy sets may be of considerable use in developing effective methods of model self-evaluation. Such methods can ultimately be used as part of a fuzzy model controller that can detect semantic errors occurring and decide when to switch to alternative models or model forms. In a system such as KBLIMS this would be another layer of intelligence that would automatically handle the simulation portion of a query. The user would largely be unaware of the specifics of this intelligent agent [29]. However, it would increase confidence in the logical consistency of the answer that the user receives from the system.

5 Discussion

Using KBLIMS as an example of an intelligent geographical information system application, it has been demonstrated that there are many aspects of such a system where fuzzy sets and fuzzy logic would be useful as way of addressing many issues of managing the uncertainty in an intelligent GIS.

5.1 Classifications and Landscape Analysis

The advantages of incorporating fuzzy sets in the terrain analysis level may not be fully exploited using contemporary approaches that stress the defuzzification of fuzzified information. Thus there is a need to be able to incorporate the uncertainty represented in these approaches in such a manner that it can be of some use to the intelligent GIS, especially any models that may be simulating physical processes. The uncertainty model proposed by Davis and Keller [12] illustrates how one may incorporate fuzzy metadata concepts with Monte Carlo simulation to eventually support a system that can use the uncertainty information to detail how much to trust the model output. The question becomes how to translate it into information that is useful for subsequent use to a system such as KBLIMS.

There is a much larger research issue that goes beyond using fuzzy sets in methods to improve upon the accuracy of classifications such as soils or vegetation. In the field of GIS there is very little that is explicitly known about the relationship between the dynamic real-world system being modeled, or simulated, the spatial sampling strategies used to collect the information, and the vagueness in a classification scheme. Results of research addressing these issues should provide valuable insights for the planning of spatial sampling activities, choice of data acquisition technologies, and our understanding of the effect that these choices have on our description of ecosystems. There is a research programme underway designed to establish explicit linkages between the level of uncertainty in a description of the dynamics of an observed ecosystem, and the spatiotemporal characteristics of the sampling strategy used to observe the ecosystem. It uses a spatial simulation model as a surrogate for a real-world ecosystem. Samples are collected from the simulation as it runs. The samples are to be collected according to different classification schemes, fuzzy and crisp. They will be compared to actual simulation behavior. Thus, the balance between sampling accuracy and fuzziness of the classification scheme will be explicitly related to the complexity of the system [17].

5.2 Object Models and Geographical Queries

At the object model level of KBLIMS is where management of uncertainty using fuzzy sets can be made most explicit and thoroughly integrated into the system. To a large extent incorporation of fuzzy instances, objects, and classes will depend to a large extent upon the nature of the data provided by the terrain analysis and the needs/capabilities of the query model. It is at this level that what has traditionally been described as two separate areas - representation and spatial analysis [30] - become integrated. In order for KBLIMS to have a working object model the behaviour of the fuzzy objects, relations, or classes would by necessity be implemented. In many cases this would be tantamount to many of the functions collectively described as spatial analysis. However, a portion of the spatial analysis functions would be incorporated into portions of the query model.

It is at the object model level where specification of fuzzy relations and classes may not have an intrinsic appeal to the scientist-user. Since the visual query system of KBLIMS exploits the object model schema and is used to compose queries it may be useful for the management of fuzzy objects, relations, and classes to be handled explicitly by KBLIMS, yet appear to the scientist-user as the usual objects and relations. This suggests that the scientist-user may be kept distant from the explicitly fuzzy set aspects of the system. And any user interaction with fuzzy objects, relations, classes is mediated by KBLIMS.Regardless of the level at which issues of managing fuzzy objects are held, there remain significant questions about how to approach the management of uncertainty in the supertype/subtype hierarchy. The two main approaches depend on whether the fuzzy supertype/subtype is explicitly specified by a membership value or whether an implicit definition is used. The approach using implicit definitions typically rely on use of predicates or constraints on the values for a set of attributes of the subtype taken from the supertype. Thus, it is not surprising that it is clear that more work need to be undertaken to in order to establish a complete, consistent, and meaningful definition for fuzzy extensions to a revised ODMG object data model [14]

However, when it comes to specifying the meaning of linguistic variables, especially those used as constraints such as *south-facing* hillslopes, it will be difficult to justify keeping the fuzzy portion of the system completely hidden from the user. A useful approach for specifying these fuzzy "concepts" is demonstrated in [22] where fuzzy concepts are formed through an automated question/answer dialogue between machine and user where only yes/no responses from the user are required. The use of this approach opens the door to the possibility of building personal versions of linguistic variables or deriving a group's fuzzy concept as a result of human-machine interaction [22]. This is more directly related to the simpler retrieval functions of the query system. More complex questions emerge when considering the management of uncertainty with regard to the simulation models used in the answering of queries regarding hydro ecological processes.

5.3 Model Integration

Geographic information systems like KBLIMS view environmental simulation models as information providers. The ability of the system to self-evaluate the integrity of a simulation model operation as simulations progress is important in performing control system functions about what parameters to adjust dynamically or which submodels to switch to when semantic error becomes unacceptable [29].

Fuzzy sets can play a significant role at many different levels in the semantic integration analysis of environmental models within an intelligent geographic information system. Semantic weaknesses can be used to develop a measure of semantic proximity that may be presented in the form of a map of fuzzy memberships. Linguistic terms used to describe essential characteristics in sensitivity analysis can be utilized in a multiple criterion decision making process based on fuzzy logic to arrive at a semantic reconciliation which is consistent with the context of a query [27,28]. This approach may enable the preservation of as much end-user and model developer autonomy as is practical. There is much empirical work to be done with existing models, integrating them into an intelligent geographic information system environment and to determine how well semantic integration can be addressed in a variety of contexts.

6. Concluding Comments

Many advanced GIS applications such as KBLIMS are not designed at the beginning to accommodate explicitly managing uncertainty that may be appropriately done so using fuzzy set theory. There are piecemeal portions of the system which may or may not exploit the rich research of fuzzy sets to address very specific problems such as model self-evaluation and control which will allow the system to increase its ability to manage uncertainty generated by its own activities. However, as the application domain evolves and experience with formalizing that domain increases, some significant kinds of uncertainty may be specified and ultimately managed with fuzzy sets. It is evident that in GIS applications, such as KBLIMS, that simply arriving at membership values to describe variations in soils and their properties over space is really the inauguration of a process of formalizing and managing uncertainty that cascades through the system warranting changes at the object model level.

References

1. A. Morris, F.E. Petry, and M. Cobb, Incorporating spatial data into the fuzzy object oriented data model, *Proceedings of Seventh International Conference on Information Processing and Management of Uncertainty in Knowledge Based Systems* (IPMU 98), New York, NY, 604-611, 1998.1]
2. A. Morris, *A Fuzzy Object Oriented Approach for Handling Spatial Data with Uncertainty,* Ph.D. dissertation, Tulane University, New Orleans, Louisiana, USA, 1999.
3. V.B. Robinson, Implications of fuzzy set theory for geographic databases, *Computers, Environment, and Urban Systems*, 12: 89-98.
4. V.B. Robinson, and D.S. Mackay, Semantic modeling for the integration of geographic information and regional hydroecological simulation management, *Computers, Environment, and Urban Systems*, 19 (5/6): 321-339, 1996.
5. D.S. Mackay, V.B. Robinson, and L.E. Band, On a knowledge-based approach to the management of geographic information systems for simulation of forested ecosystems, In: Michener, J. and J. Brunt (eds) *Environmental Information Management and Analysis: Ecosystem to Global Scales*, Taylor & Francis Publishers, 321-339, 1994
6. V.B. Robinson, and D.S. Mackay, knowledgebased land information manager and simulator (kblims) for forested ecosystem simulation management, *IJCAI-95 Workshop on Artificial Intelligence and the Environment*, International Joint Conferences on Artificial Intelligence, Montreal, Canada, August, 100-108, 1995.
7. R.A. MacMillan, W.W. Pettapiece, T.W. Goddard, and S. Nolan, A generic procedure for automatically segmenting landforms into landform elements using

DEMs, heuristic rules, and fuzzy logic, *International Journal of Fuzzy Sets and Systems*, (in press)

8. G.M. Foody, Fuzzy modelling of vegetation from remotely sensed imagery, *Ecological Modelling*, 85: pp.3-12, 1996.
9. A-Xing Zhu,. A similarity model for representing soil spatial information, *Geoderma*: 77: 217-242,1997.
10. A-Xing Zhu, A personal construct-based knowledge acquisition process for natural resource mapping, *International Journal of Geographical Information Science*, 13(2): 119-141, 1999.
11. M.A. Cobb, F.E. Petry, and K.B. Shaw, Extensions to geometric approximations of spatial boundaries and assessment of fuzzy spatial relations, *International Journal of Fuzzy Sets and Systems*, 1999(in press).
12. T.J. Davis, and C. P. Keller, Modelling uncertainty in natural resource analysis using fuzzy sets and Monte Carlo simulation: slope stability prediction, *International Journal of Geographical Information Systems*, 11(5): 409-434, 1997.
13. F. Wang and G.B Hall,. Fuzzy representation of geographical boundaries in GIS, *International Journal of Geographical Information Systems*, 10(5): 573-590, 1996.
14. V. Cross, and A. Firat. Fuzzy objects for geographical information systems, *International Journal of Fuzzy Sets and Systems*, 1999(in press).
15. T. Hadzilacos and N. Tryfona, 1996. Logical data modelling for geographical applications, *International Journal of Geographical Information Systems*, 10(2), Taylor & Francis, London 1996, pp.179-203.
16. G. Vert, A. Morris, M. Stock, P. Jankowski, Extending ERD modeling notation to fuzzy management of GIS datasets, *Proceedings of 18th International Conference of the North American Fuzzy Information Processing Society* (NAFIPS 99), IEEE, New York, 819-823, 1999.
17. P. A. Graniero, 1999. The effect of spatiotemporal sampling strategies and data acquisition accuracy on the characterization of a dynamic ecological system, , *Proceedings of 18th International Conference of the North American Fuzzy Information Processing Society* (NAFIPS 99), IEEE, New York, pp.829-833, 1999.
18. K. Tanaka, S. Kobayashi and T. Sakanoue, Uncertainty management in object-oriented database systems, *Proceedings* Database and Expert System Applications, 251-256, 1991.
19. F. Wang, Towards a natural language user interface: an approach of fuzzy query, *International Journal of Geographical Information Systems*, 8(2): 143-162, 1994.
20. F. Wu, A linguistic cellular automata simulation approach for sustainable land development in a fast growing region, *Computers, Environment, and Urban Systems*, 20(6): 367-388, 1996.
21. V.B. Robinson, and D.S. Mackay, Intelligent visual query in a knowledge-based coupling of geographic information and ecosystem simulation, *Proceedings* GIS/LIS'96, 39-44, 1996.
22. V.B. Robinson, Individual and multiperson concepts of spatial relations acquired through human-machine interaction, *International Journal of Fuzzy Sets and*

Systems. (In press).
23. F. Wang, A fuzzy grammar and possibility theory based natural language user interface for spatial queries, *International Journal of Fuzzy Sets and Systems*, (in press).
24. J. Kacprzyk, S. Zadronzny, FQUERY for ACCESS: fuzzy querying for a Windows-based DBMS, In P. Bosc and J. Kacprzyk (eds), *Fuzziness in Database Management Systems* , Physica-Verlag, 415-433, 1995.
25. H. Couclelis, Towards an operational typology of geographic entities with ill-defined boundaries, In P. Burrough and A. Frank (eds), *Geographic objects with indeterminate boundaries*, GISDATA Series, Taylor & Francis, London, UK, 45-56, 1996.
26. V. Gaede and O. Gunther, Multidimensional access methods, *ACM Comp. Surveys*, 30(2): 170-231, 1998.
27. D.S. Mackay, Semantic integration of environmental models for application to global information systems and decision-making, *ACM SIGMOD Record*, 28(1): 13-19, 1999.
28. D.S. Mackay and V.B. Robinson, A multiple criteria decision support system for testing integrated environmental models. *International Journal of Fuzzy Sets and Systems*. (in press).
29. D.S. Mackay, Integrating self-evaluating hydrological and ecological models of different spatial scales, *Proceedings* GIS/LIS'97, Am. Soc. Photogrammetry & Remote Sensing, Bethesda, MD, p. 486-498, 1997.
30. D. Altman, Fuzzy set theoretic approaches for handling imprecision in spatial analysis, *International Journal of Geographical Information Systems*, 8 (3): 409-434, 1994.

Acknowledgements

This work was partially supported by an operating grant from the Natural Sciences and Engineering Research Council (NSERC) of Canada. Work on KBLIMS was partially supported by operating grants from NSERC, a NSERC Doctoral Fellowship and Tri-Council EcoResearch Doctoral Fellowship. The author is grateful to Phil Graneiro and an anonymous reviewer for comments that substantially improved the quality of the manuscript.

Conceptual Modeling of Geographic Information System Applications

Adnan YAZICI[1] and Kemal AKKAYA[1]

[1] Department of Computer Engineering, Middle East Technical University, 06531, Ankara / Turkey, e-mail: {yazici,akkaya}@ceng.metu.edu.tr

Abstract. An important research trend in databases is to handle different types of uncertainty at both conceptual and logical levels for various non-traditional applications that may involve imprecision and uncertainty that have been difficult to integrate cohesively in simple database models. In this study we describe how to conceptually model complex and uncertain information at the conceptual level (by utilizing the $ExIFO_2$ model) for Geographic Information System (GIS) applications. We also give the mapping algorithm that transforms the conceptual schema of GIS applications into the logical database structures by utilizing the fuzzy object-oriented database (FOOD) model at the logical level. The types of uncertainty that we mainly focus in this paper are fuzzy, null, and incomplete information related to objects and their properties, classes, and relationships among objects and classes of GIS applications.

Keywords. Uncertainty, Fuzziness, Object-Oriented Databases, Conceptual Modeling, and Geographic Applications

1 Introduction

In general, a conceptual model is a type of abstraction that uses logical concepts and hides the details of implementation and data storage. Some of the existing conceptual models offer powerful concepts to the designers that provide getting the most complete specification from the real world [1,2,3,4]. However, the classical models in this group generally suffer from the lack of concepts (object-identity, reusability, representation of imprecise information, etc.) which are important for modeling various complex applications such as Multimedia Database Systems (MMDBS), Office Automation Systems (OAS), Geographic Information Systems (GIS), Expert Database Systems (EDS) and so on. Modeling requirements for these complex applications and flaws in the currently used models have led to the definition of more powerful conceptual models such as the $ExIFO_2$ model [5,6] and the others [7]. However, there still have been a number of research issues related to conceptual modeling with increasing complexity in the complex applications. As some researchers [1,2,4,8] have pointed out, the general-purpose conceptual models are not adequate for GIS applications. Therefore, there is a need for utilizing a powerful conceptual model to satisfy the unique requirements of these applications, such as the specific properties of geographic objects or entities and relationships, which may involve various forms of uncertainty.

There are two common data models for modeling spatial information in GIS applications: field-based models and object-based models [8,9,10]. Field-based models treat the spatial information space as a continuous domain such as altitude, rainfall and temperature - as a collection of spatial functions transforming a space-partition to an attribute domain. The object-based models treat the information space as if it is populated with recognizable objects (spatial and aspatial) that are discrete and spatially referenced. The modeling approach that we take in this study is object-based. That is, the database for GIS applications stores a map that consists of a collection of identifiable objects, which refer to the partitions and fragments of information space. Geographical objects have geometric properties that can be modeled by the measurements, properties, and relationships of points, lines, and polygons. Some geographic objects may also have topological properties.

There have been previous works related to conceptual modeling of geographic applications by either introducing a new conceptual model or extending one of the existing models, mainly the Enhanced Entity-Relationship (EER) model. GraphDB [11], GODOT [12], and the system by Worboy [13] are some of those that attempt to model GIS applications using an object-based approach. GISER [8] and Geo-ER [2], which are extensions to the ER model, are some of recent attempts that try to unify continuous fields with entities and relationships. The Geo-ER model provides a set of concepts as an add-on to the ER model and attempts to capture spatial peculiarities at the conceptual level of geographic database design. The GISER model is based on four major concepts. These concepts are: space/time (boundless multidimensional extends in which geographic phenomena and events can occur and have relative position and direction); feature (geographic phenomena such as cities, mountains, etc.); coverage (a set of spatial objects); and spatial object (a subset of space and time.) All of these conceptual models assume that GIS applications do not involve any form of uncertainty.

However, it is not always possible to describe all the semantics of real world information precisely since the observation and capturing of some real world objects are not perfect, hence its modeling and representation are deficient. Uncertainty might arise from the data itself and/or the relations between objects. As a consequence of this, queries involving imprecise and uncertain information may be unavoidable. The management of uncertainty has been ignored for a long time, but it is now inevitable to be able to incorporate uncertainty in many complex applications including GIS applications. More specifically, the need for incorporating uncertainty in a conceptual modeling of GIS applications arises from the following reasons: (1) Some geographic information in GIS applications is inherently imprecise or fuzzy. For example, locations of geographic objects, some of spatial relationships, and various geometric and topological properties usually involve various forms of uncertainty. (2) Only few natural geographic phenomena have boundaries that can be accurately represented as mathematical lines. For example, vegetation types, soils, slopes, wildlife habitats, etc., all have fuzzy boundaries due to the transitional nature of variation in the phenomenon. (3) Some measurements related to spatial domain are often incomplete. Sometimes forcing such data to be completely crisp may result in falsity and useless information. (4) In GIS applications, some of the spatial domain related knowledge is specified in natural language by using fuzzy terms and numerous quantifiers (e.g., many, few, some, almost, etc.). Most of these quantifiers are fuzzy and are used when conveying vague information.

Since almost none of the published studies modeling GIS applications deals with uncertain information at the conceptual level, there is a need not only for dealing with spatial objects, but also for handling uncertain properties including inherently fuzzy ones. In addition, transforming the conceptual schema into the logical one (the object-oriented database (FOOD) model is utilized in this study) should be done in a way that it is straightforward and preserves most information.

In this paper a conceptual model, called the ExIFO_2 model, is utilized to conceptually model geographic applications. This model includes representation of spatial objects along with their geometric and/or topological properties in addition to uncertain ones. The conceptual ExIFO_2 model attempts to preserve the acquired strengths of semantic approaches, while integrating concepts of the object-oriented paradigm and uncertainty.

The next section includes a brief description of the ExIFO_2 model [5,6] and the fuzzy object-oriented database (FOOD) model [6,14]. In Section 3 we present conceptual modeling of GIS applications. Before the conclusions in the last section, we present the mapping algorithm that transforms the conceptual schema represented with the ExIFO_2 model into a logical database description by utilizing the fuzzy object-oriented database (FOOD) model.

2 Background

In this section, we discuss the basic definitions and characteristics of the models and concepts used. Topics include a brief background on the ExIFO_2 model and the FOOD database model.

2.1 The ExIFO_2 Model

The ExIFO_2 model is an extension of the combination of the semantic data models IFO_2[15] and ExIFO[5,6], which are each an extension of the IFO model [1]. The ExIFO_2 model is a mathematically defined data model that incorporates the fundamental principles of semantic database modeling within a graph-based representational framework. More formally, an ExIFO_2 schema is a directed graph with various types of vertices and edges, representing atomic objects, constructed objects, fragments and ISA relationships.

Two types of uncertainty are dealt with in the ExIFO_2 model. The first type of uncertainty is considered to be at the *attribute level*, meaning that some attributes of objects may have uncertain values. The second type of uncertainty is considered to be at the *object* and *class level*. That is, some objects may have instances whose membership to the object set may be graded in [0,1] (i.e., partial membership of an instance of an object type) and/or some subclasses whose membership to the superclasses (class/subclass level uncertainty) may be in [0,1]. An explicit definition of the object identifier, which is object value independent, is integrated in the model. To achieve these, all manipulated elements of the ExIFO_2 model are redefined. On the other hand, integrating the concepts of *alternative, composition* and *grouping* for complex objects enhances the modeling power of the ExIFO_2 model.

Graphical representation and a brief description of the related concepts of the $ExIFO_2$ model are included below. All types of the $ExIFO_2$ model and constructors are shown graphically in Figure 1, Figure 2 and Figure 3.

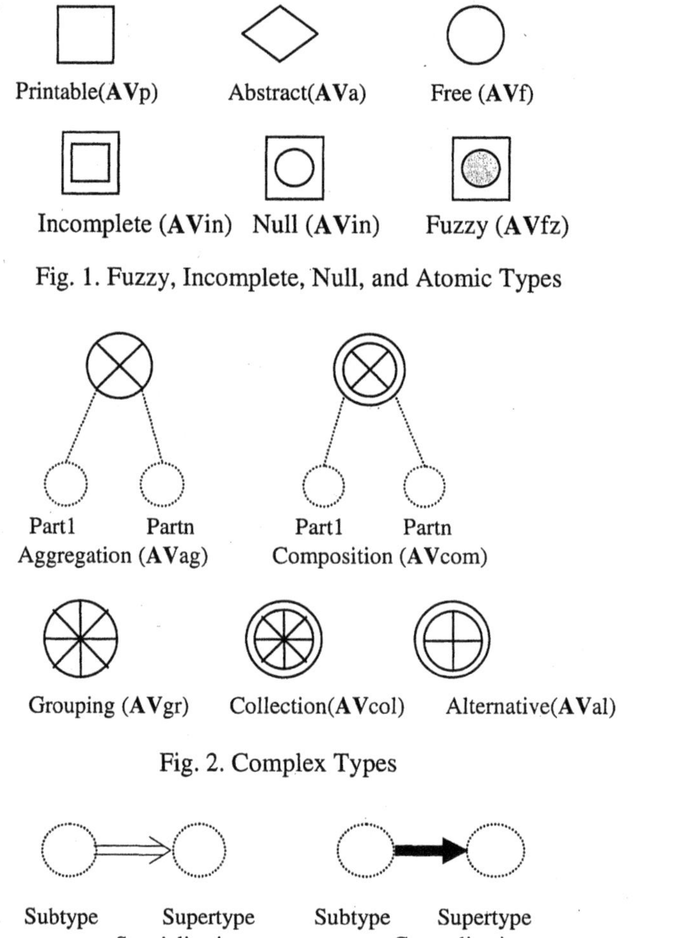

Fig. 1. Fuzzy, Incomplete, Null, and Atomic Types

Fig. 2. Complex Types

Fig. 3. Function Types and ISA Links

We consider three kinds of uncertainty in the $ExIFO_2$ model:

The true data may belong to a specific set of values, *incompleteness (or imprecision)* [16],

The true data value is not known, *null*,

The true data is available, but specified only with a descriptive term, *fuzzy* [17].

Uncertainty is also considered at two levels: *Attribute* level and *object-class* level. For the representation of attribute level uncertainty in the $ExIFO_2$ model, three constructors (*fuzzyset, incompleteset*, and *null* constructors) are defined. Using these constructors it is possible to explicitly represent attributes having uncertain values. The second level of uncertainty considered is the fuzziness of instances of specific objects in the corresponding object set (class/object level) and fuzziness of subclasses in superclasses (class/subclass level). In class/subclass level "F" will be used on generalization and specialization arrows in the representation of the object, to indicate the possibility of gradual membership. (e.g. a *caravan* class can be considered as a subclass of *house* class with a membership degree of 0.6).

There are two main groups of types in the $ExIFO_2$ model. The first group are the atomic types (*printable, abstract* and *free*) and the second group are the complex types (*aggregation, composition, collection, grouping, union, fuzzySet, incomplete, and null* costructors) produced by the use of constructors.

The *printable* objects correspond to objects of predefined types that serve as the basis for input and output. The *abstract* objects correspond to objects in the world that have no underlying structure. The third type of atomic objects is *free*, and corresponds to entities obtained via the *'ISA'* relationship. The graphical representations of these atomic types are given in Figure 3. Non-atomic (or constructed) types are formed from an underlying type by applying a finite set of constructors. These constructors (except *fuzzyset, incompleteset* and *null* constructors) may be applied recursively to build more complex types.

The *grouping* constructor is a high-level mechanism of the $ExIFO_2$ model and is an abstraction in which the relationship among elements is considered as a set of objects. This constructor depicts the type corresponding to sets of data values of an attribute. The *grouping* constructor has 'AND' semantics, since each member of the set necessarily and precisely belongs to the set. Hence, this type of the constructors can capture the attributes that are both multi-valued and crisp. All objects in the *grouping* abstraction are of the same type and not ordered. An example for this constructor is the set of *authors* of a book. A set of authors of a specific book is precisely known and crisp; therefore, the authors are related with '*AND*' semantics. That is, the names of the authors specified are the only authors of the book, not some subset of the names of that set.

The *collection* types include an exclusivity constraint for the *grouping* constructor and this is the only difference between the *collection* and *grouping* constructors.

The *aggregation* constructor represents the *aggregation* abstraction of semantic models defined by the Cartesian product. This constructor connects a subtype representing a part of an object to the type representing the entire object; thus, building a higher level object. This abstraction ignores some individual differences of the aggregated types. For example, object type *motor-boat* is viewed as being an ordered pair of *hull* and *motor*. Semantically, the *aggregation* constructor derives only objects fully supported by the type, but, unlike the *grouping* constructor, the objects are ordered. An example is the list of components of an address of a person (i.e., *street-name, zip code, city*, etc.).

The *composition* constructor also represents the *aggregation* abstraction of semantic models defined by the Cartesian product. The only difference of this constructor from the *aggregation* constructor is that it provides an exclusive structure.

In the $ExIFO_2$ model, structurally different types are handled by using the *alternative* type concept. This constructor represents the IS_A generalization link enhanced with a disjunction constraint between the generalized types.

With a *fuzzyset* constructor, a set of values of a given specified domain of a type (attribute in relational model terminology) can be defined. Only a subset of these values that corresponds to the object of that structure type are true instances. This constructor builds an instance in the form of a fuzzy set whose elements are related to each other fuzzily. That is, any instance of the set belongs to the set with a degree in [0,1] and all of the elements are said to have similarity relationship to each other to that degree[18].

The *incompleteset* constructor is used to represent *incomplete* information, such as range values. Semantically, an object is the only instance of the corresponding type. This constructor represents a given specified domain of a type, which the constructor defines, but it involves different semantics compared to the *fuzzyset* constructor. Only one of the values from a specified range corresponds to an instance of that structure type. Therefore, the *incompleteset* constructor has 'XOR' semantics, since only one instance of the set represented as a range is the true instance of the underlying type.

Another kind of constructor for representing uncertainty is the *null* constructor. This constructor is used to represent attributes whose domain is extended to handle various types of null values. The null values taken into consideration in the $ExIFO_2$ model are as follows: the true value is not known (*unk*), does not exist (*dne*), or there is no information (*ni*) on whether a value exists or not. All of these can be considered as a type of a null value and depending on the situation, different interpretations given to the null value. For example, if we know that a person does not have a telephone, then it can be evaluated as *dne*. If we know the person has a telephone but we do not know the number, then it is *unk*. If we do not know whether or not a telephone exists, then it is *ni*.

The types in the $ExIFO_2$ model can be linked by using functions (called *fragments*). The goal of a fragment is to describe the properties of the principal types. Functions can be (simple, complex -multivalued-) or (partial-0:N link-, total-1:N link-). In a partial function some elements of the domain have no associated elements in the codomain, otherwise it is called a total function [11]. Any total or partial function can be complex or simple.

A set of schema invariants and update rules that consists of the explicit statement of some rules and constraints implicit in the model are defined. Additionally, some of the restrictions concerning the fuzzy and object-oriented extensions are also included in the model. In order to provide a verification of an $ExIFO_2$ schema while preserving a consistent representation, a set of update rules are defined. These rules are used especially for modifying an existing $ExIFO_2$ schema. Whenever a change is done to the schema, these rules are checked to see if the schema invariant constraints are violated or not. Here we do not describe all of the details of the $ExIFO_2$ model due to space limitations. The details (along with a number of examples) of the model can be found in [5,6].

2.2 The Fuzzy Object-Oriented Data (FOOD) Model

Here we briefly review the fuzzy object-oriented data model (the FOOD model) [5,14], since the FOOD model is utilized as the logical model into which to map the conceptual schema of a GIS application. The basis of the fuzzy object-oriented database model is the similarity relation. For each fuzzy attribute, a fuzzy domain and a similarity matrix are defined. Similarity matrices are used to represent the relation within the fuzzy attributes. The domain, *dom*, is the set of values the attribute may take, irrespective of the class it falls into. The range of an attribute, *rng*, is the set of allowed values that a member of a class, i.e. an object, may take for an attribute. In general $rng \subseteq dom$. For instance, assume that *height* is an attribute and the domain of *height* is between 0 and 230 cm. If there exists a *Student* class, the range of the *height* attribute for this class may be 130cm to 230cm. A range for each attribute of the class is defined as a subset of a fuzzy domain. The range definition for attribute a_i of class C is represented by the notation, $rng_C(a_i)$, where $a_i \in Attr(C) = \{a_1, a_2, ..., a_n\}$. *Attr(C)* refers to the attributes of class C. Similar objects are grouped together to form a class and fuzziness at object/class and class/superclass levels are represented this way. The idea of fuzziness extends in the relation of an object with the class of which is created as an instance. An object belongs to a class with a degree of membership. Based on the considerations of relevance and ranges of attribute values, the membership of object o_j in class C with attributes *Attr(C)* can be defined as

$$\mu_C(o_j) = g[f(RLV(a_i,C), INC(rng_C(a_i)/o_j(a_i)))]$$

where $RLV(a_i,C)$ indicates the relevance of attribute a_i to class C, and $INC(rng_C(a_i)/o_j(a_i))$ denotes the degree of inclusion of the attribute values of o_j in the formal range of a_i in class C. The degree of inclusion, determines the extent of similarity between a value (or a set of values) in the denominator with a value (or a set of values) in the numerator. Function f represents the aggregation over n attributes in the class and g reflects the type of link existing between an instance (object) and a class/superclass (f and g are functions that may be inherited from the superclass or may be defined within the local class). The value of $RLV(a_i,C)$ may be supplied by the user or computed. Several cases (which we will not discuss here) are possible for the evaluation of $INC(rng(a_i)/o_j(a_i))$ [5].

Fuzziness may occur at three different levels in our fuzzy object-oriented database model; the attribute level, the object/class level and the class / superclass level. We will only summarize some of the basic concepts of the model here, a detailed description along with the examples of how fuzziness is handled in each level in the model is given in [5,14].

Attribute Definitions

Unlike the fuzzy relational model, in the fuzzy object-oriented model, attributes can have a set of values (leading to multivalued attributes) connected with a logical operator AND/OR/XOR. The attribute value sets are differentiated according to their semantics. The following notation is used to indicate AND, OR or XOR multivalued attributes:

logical operator: AND:<.......>, OR:{..... } and XOR:[......]

Assume that the domain of foreign languages is = *{English, German, French, Italian, Spanish, Dutch, Turkish, Chinese, Japanese, Russian}*. The following interpretations are valid:

Mathew.Lang = <English, German>,
i.e., "Mathew can speak English and German".
Helga.Lang = {German, French},
i.e., "Helga can speak German or French or maybe both."
Hazal.Lang = [Italian, Turkish],
i.e., "Hazal can speak one language either Italian or Turkish.".

Class Definitions

Every class has a range definition for each of the fuzzy attributes with the corresponding relevance rules indicating the importance of that attribute in the definition of that class. In this way an "approximate" description of the class is given. An attribute of a class is allowed to take any value from the domain without considering the range values.

In the FOOD model, semantics is associated with the range definitions to permit a more precise definition of a class. The set given in the similarity-based range definition includes the semantics *OR, AND,* or *XOR*. *AND* semantics forces a multivalued use of the attribute and *XOR* forces a singleton attribute. *OR* is the most uncertain definition that can be made for a class definition. The logical relation is determined in the range definition, and the instances with multivalued attributes obey this relation. For instance class *C* having attributes *a, b, c* from domains *A*, *B*, and *C* respectively can be defined as:

Class C:

$rng_C(a) = \{a_1, a_3, a_6\}$	$dom_C(a) = \{a_1, a_2, a_3, \ldots, a_k\}$
$rng_C(b) = <b_5, b_7>$	$dom_C(b) = \{b_1, b_2, b_3, \ldots, b_m\}$
$rng_C(c) = [c_2, c_4, c_6]$	$dom_C(c) = \{c_1, c_2, c_3, \ldots, c_n\}$

Class *C* has "a_1, a_3 or a_6" values for attribute *a*, "b_5 and b_7" requiring multivalued use of attribute *b*, and finally a value of either "c_2", "c_4", or "c_6" for attribute *c*, only one of which is true. This is a special case of *XOR*; it is true only when exactly one of the entries is valid.

If the range definition of attribute *age* of class *Person* was given *as rng(age) = {young, very young}*, then the objects use the same logical operator (*OR*) for attribute *age*:

$o_1(age) = \{very\ old, old\}$, $o_2(age) = \{very\ old, young, old\}$

Relevance weights are assigned for each attribute, and they show the significance of the range definition of that attribute on the class definition. If relevance rules for class *C* are given as:

$RLV(a) = 1.5$, $RLV(b) = 2.5$, $RLV(c) = 0.2$

Then attribute *b* is the most important attribute defining class *C*, attribute *a* more or less determines class *C*, and attribute *c* is of very little importance in determining class *C*.

Object/ Class Relations

The object/class level denotes the membership degree of an object to a class. The main feature that distinguishes fuzzy classes from crisp classes is that the boundaries of fuzzy classes are imprecise. The imprecision of the attribute values causes imprecision in the class boundaries. Some objects are full members of a fuzzy class with membership degree 1, but some objects may be related to this class with a degree between 0 and 1. In this case they may still be considered as instances of this class with the specified degree in [0,1]. In the FOOD model a formal range definition indicating the ideal values for a fuzzy attribute is given in the class definition. However, an attribute of an object can take any value from the related domain. So, the membership degree of an object to the class is calculated using the similarities between the attribute values and the class range values, and the relevance of fuzzy attributes. The relevance denotes the weight of the fuzzy attribute in the determination of the boundary of a fuzzy class. If an object has the ideal values for each fuzzy attribute, then this object is an instance of that class with a membership degree of 1. Otherwise, it is either an instance with a membership degree less than 1, or it is not an instance at all (when the membership degree is smaller than the threshold value) depending on the similarities between attribute values and formal range values. The closer the attribute values to the range, the higher the membership degree of the object. If an attribute value of an object is crisp, the membership degree of this crisp value to the fuzzy terms in formal range definition is calculated and used to find the object membership degree to the class. The system calculates the membership degree of objects to their classes during object creation and updating by using the formulas given below with the related semantics.

To calculate the membership degree of an object to a class, we calculate the inclusion degrees of attribute values in the range of attributes. Since the attribute values may be connected through *AND, OR*, or *XOR* semantics, the inclusion value depends on the attribute semantics.

The more similar an object's attribute value to the range definitions, the higher the class/object membership degree. But how is this inclusion determined? The membership degree of object o_j to class C is determined using:

$$\mu_C(o_j) = \frac{\sum INC(rng_C(a_i) / o_j(a_i)) * RLV(a_i, C)}{\sum RLV(a_i, C)}$$

where $INC(rng_C(a_i)/o_j(a_i))$ is the value of the inclusion taking into account the semantics of multivalued attribute values (as will be described below), $RLV(a_i,C)$ is the relevance of attribute a_i to class C and is given in the class definition. The weighted-average is used to calculate the membership degree of objects. All attributes, therefore, affect the membership degree proportional to their relevance. The formulas used to calculate inclusion degrees for the *AND, OR*, and *XOR* connection semantics are given in [5,14].

Class/Subclass Relation

The next importance relationship is between a class and superclass, in other words the answer to the question "To what extent the class belongs to its superclass". The FOOD model is the basis of the formulation to find the degree of membership of a class in its superclass.

In the FOOD model, an approximation approach is used to find the inclusion degree of a class in a superclass. Consider class *Vehicle* defined as having a body and a motor and seats, represented by aggregating these parts as *<body, motor, seat>*. A tree which has a body (trunk) is much different from a vehicle, so is a theater saloon with seats. The similarity-based model assumes that the class/subclass relation is built logically at the beginning and such a class/subclass relation will probably not be built in the database. However, it is always preferable that the model accounts for arbitrary modeling. We do not include the details here, but the interested readers can find the details of the FOOD model in [5,14].

3 Conceptual Modeling of Geographic Information Systems (GIS) Applications

A GIS application usually consists of a set of maps, which include a set of geographic objects [8,9,10,11,12,14,19]. We assume that the map is modeled here as an abstract type with two basic attributes *name, scale*, i.e. the map of *East Black Sea*, with the scale *1/5000*, and a set of geographic objects. Figure 4 shows abstract type *Map* and its attributes.

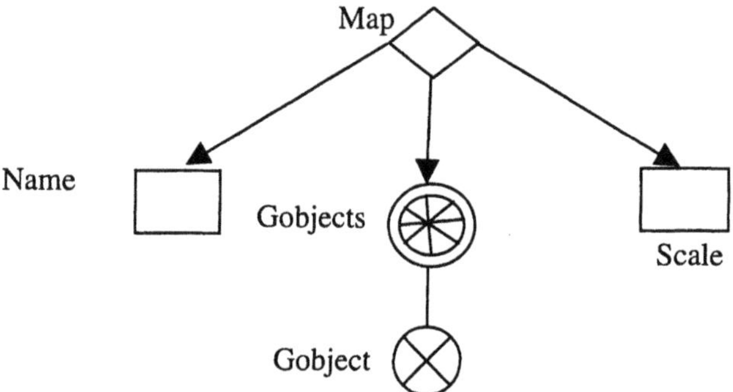

Fig. 4. The *Map* Abstract Type

Geographic objects (*Gobjects*) are a list of the *Gobject* type. Type *Gobjects* is modeled by using a *collection* constructor to group the *Gobject* objects, but being mutually exclusive, because we want to have the uniqueness constraint on the objects that it represents. *Gobject* may have three different properties. They are *GraphProp*, *Description*, and *Geos*. These are aggregated to form type *Gobject* by using an *aggregation* constructor. This is intuitive because in real-life maps, for instance, the streams of a country also may have a number of attributes.

More specifically, some of graphical properties for displaying maps include *color, lnwidth*, ant the others, depending on the application. Descriptive properties may include *length, current speed, etc.,* of each stream. An example for the spatial properties that represent the structural geometry of the streams is *lines* for streams.

Gobject may include all these properties. Figure 5 shows how to model a geographic object in the $ExIFO_2$ model by including only some of its properties.

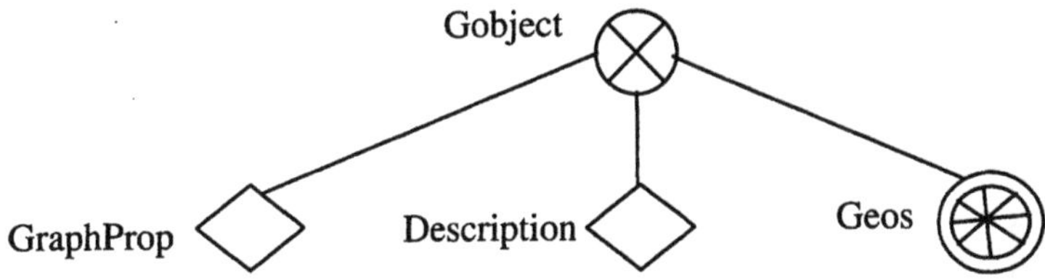

Fig. 5. *Gobject* Type in GIS Applications.

Graphical properties refer to the displaying of these objects, for instance, *color, texture, font, text, histogram* or *linewidth* of the objects. Therefore we include an abstract graphical type which consists of these attributes. The color of an object on the map can be fuzzy valued such as *dark blue, almost green,* etc. The texture may also be fuzzy valued, such as *rectangles, circular lines,* etc. *Linewidth, font, histogram* (the density of the colors in the object) and *text* are normal printable types. Figure 6 shows the modeling of *GraphProp* in the $ExIFO_2$ model.

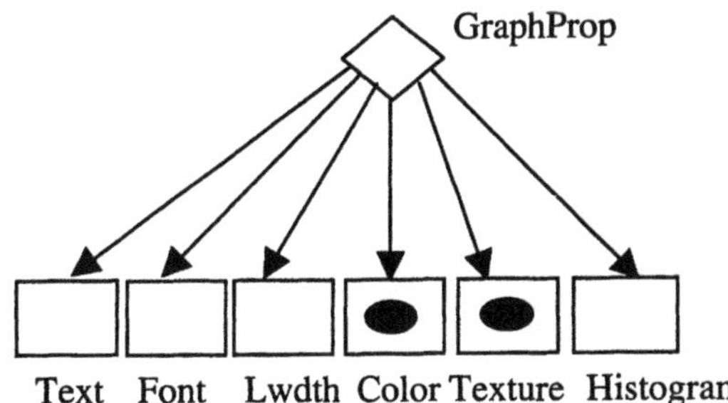

Fig. 6. Graphical Properties of *Gobject.*

The *Description* type of *Gobject* is optional, because, as descriptive information, spatial information is usually sufficient. Because the descriptive information is optional for *Gobject,* we have a partial function between *Description* and *Aspatials (non-spatial)* meaning that sometimes there may not exist description. *Aspatials* is modeled as a type that is a grouping of *Aspatial.* Here we used a *grouping* constructor, because an object may have a number of descriptive entries. For instance, for cities (a city is an object of type *region*), we have at the same time, *name*(*StrAtr*), *population*(*FuzzAtr*), *lattitude*(*FuzzAtr*) and *area*(*IntAtr*). To handle this situation, the description type is modeled by a *grouping* type, which is shown in Figure 7.

Aspatial types [12] must have names, such as *population, soil, vegetation* etc. and there must exist some values as instanses of these types. In mosts cases there are three subclasses inherited from type *Aspatial.* These are *IntAtr, StrAtr* and *FuzzAttr.* An example for *IntAtr* may be *name* of a city, an example for *FuzzAttr* may be the *population* of a city. Other than these values we sometimes may also need incomplete values such as *latitude,* specified as a range 36-42. In order to handle such types, we use the *alternative* constructor which enables us to use either the crisp value for an attribute or the incomplete value, but not both. *StrAttr* is the same as *IntAttr*. For instance, the type of *landuse* is represented by *StrAttr*. *Landuse* is the *name* for descriptive information and *urban, agriculture, brushland, forest, water, barren* are values for *landuse(StrAtr)*.

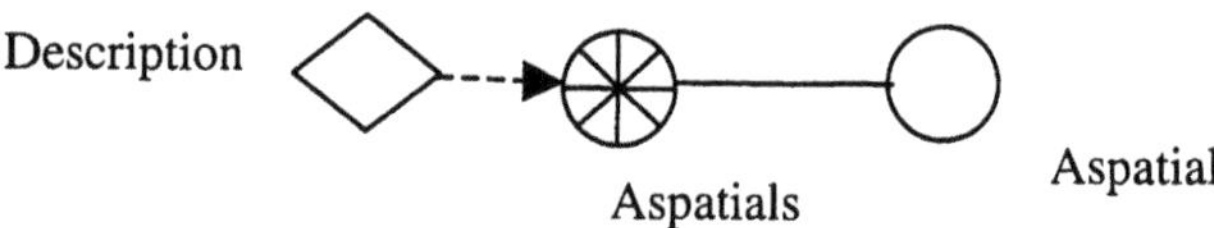

Fig. 7. The *Description* Abstract Type

A geographic object must have some spatial properties, because actual maps consist of many geometric objects. Spatial properties of *Gobject* are the geometric structure of the objects in the map, (*lines*, *regions*, *points* etc.). In the ExIFO_2 model we represent the spatial properties by a *collection* type, *Geos*, which handles all the geometrical properties for an object. The reason for using the *collection* constructor is that the Geo objects are mutually exclusive. The cities and the roads between these cities can be modeled by using *Geos* objects, which contains *regions, lines* and *points. Regions* denote the boundaries of cities, *lines* denote the roads, and *points* denote small cities. *Geo* objects are instances of the *Geos* type, which is shown in Figure 9.

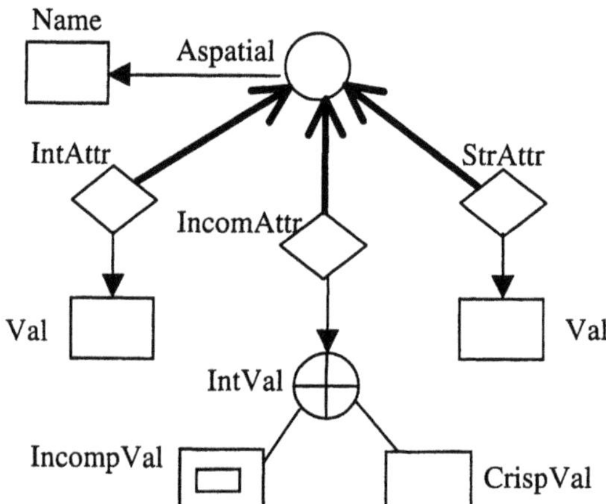

Fig. 8. The *Aspatial* Free Type

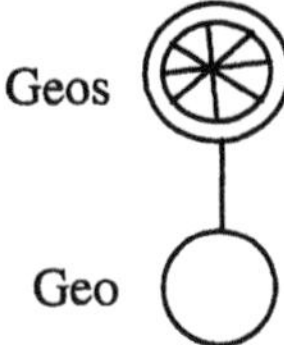

Fig. 9. The *Geos* Type with *Collection* Constructor

Geo is a generalization of *Region, Line* and *Point.* For modeling single objects, the fundamental abstractions are *point, line* and *region. Point* represents the geometric aspect of an object for which only its location in space, but not its extent, is relevant. For example a city may be modeled as a *point* in a map. A *line* is a basic abstraction for roads, rivers, cables for phone, electricity etc. A *region* is an abstraction for an object having 2d-space, e.g. a country, a lake or a national park. There are no objects other than those in *Geo*, so there are generalization links from these to the *Geo* type. Figure 10 illustrates the structure of a *Geo* object.

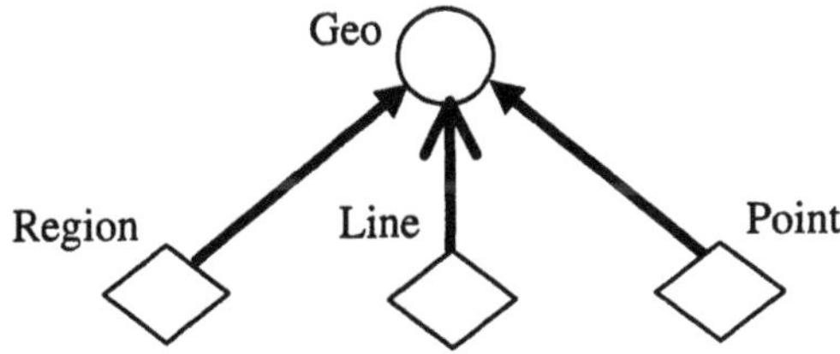

Fig. 10. The *Geo* Free Type and its Subclasses

Region is the basic type with attribute *regid.* There is a relationship between *region* and *line*, because every region has more than one line. This is modeled by complex total function between the *region* and *line* types in the $ExIFO_2$ model. Conversely, every *line* may belong to zero or more regions. Therefore we use a partial complex function between type *line* and type *region.* These relationships are shown in Figure 11.

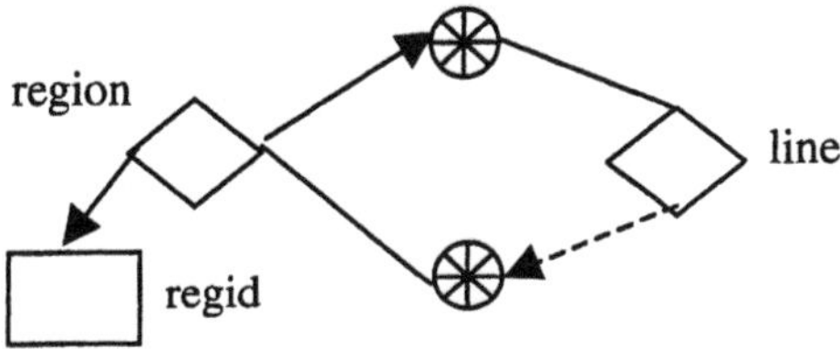

Fig. 11. The *Region* and *Line* Abstract Type and its Relationships

Some regions may have holes in them (lakes in a map). So we have another type *regionWithHoles* which is inherited from *region.* It has attribute *numofholes*, which represents the number of holes in a region. This is shown in Figure 12.

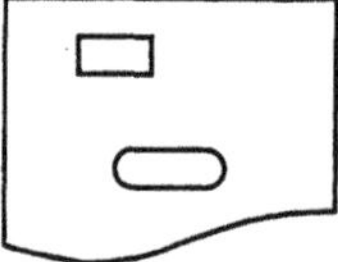

Fig. 12. An Example for *RegionWith Holes* Concept

Abstract type *line* has an attribute *lineid* and it has a construct *BegEnd.* The *BeginEnd* construct consists of two *points*: one is for the beginning of the line and the other is the end of the line. These *begin* and *end* types are of type *point*, so there is a ISA relationship between these types and type *point*, this is resulted from the specialization process (See Figure 13).

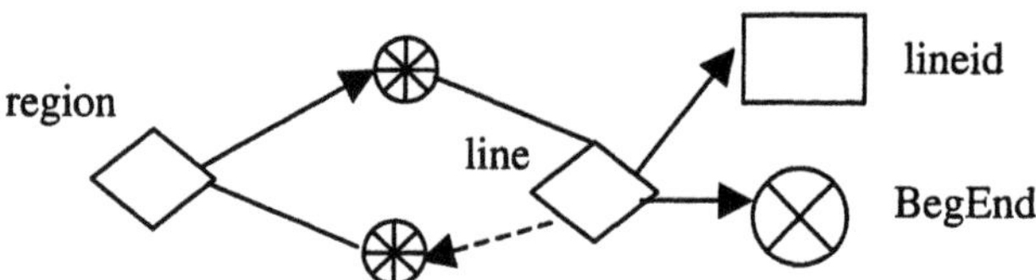

Fig. 13. The *Line* Type and its Relationship with the *Region* Type

Point is one of the basic constructs of type *GeoCons*. It has a *pointid* attribute and the *X* and *Y* coordinates. The *Coord* type is the aggregation of *X* and *Y*, as shown in Figure 14.

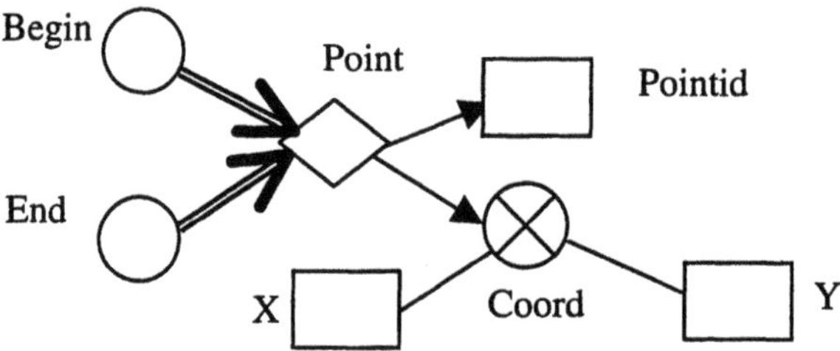

Fig. 14. The Point Type with its Related Types

Figure 15 shows the overall modeling of the GIS applications by utilizing the $ExIFO_2$ model. In the following section we outline the mapping algorithm.

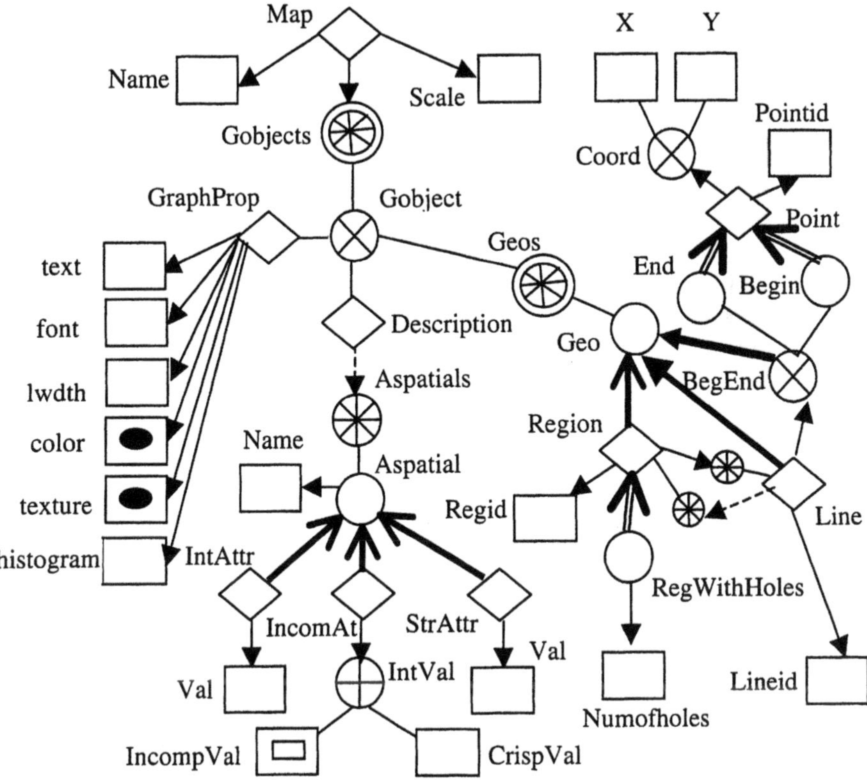

Fig. 15. Conceptual Schema of GIS Applications

4 The Mapping Algorithm

The mapping algorithm presented here transforms the ExIFO$_2$ conceptual schema into the FOOD logical database model. The algorithm is straightforward and preserves most information represented by the conceptual model.

The mapping algorithm consists of transforming the constructors of the ExIFO$_2$ conceptual model into the FOOD concepts. We first show the mapping algorithm for mapping these constructors, which include aggregation, composition, collection, alternative, and sequence, below:

```
If constructor is aggregation or composition
  create the class name
  create the attributes of the class
  if attributes of type multivalued, incompletevalued or nullvalued exist
    inherit from class FUZZY
    for multivalued, incompletevalued, or nullvalued attributes (if exists)
            create the type definition for ranges and relevance
```

```
    in the class constructor
                define the ranges, relevances, semantics
  if constructor is composition
                add a method to check exclusivity

  if constructor is  grouping or  collection
   create the class name
   create the attribute of the class
   if attribute is of type  multivalued, incompletevalued or nullvalued
       inherit from class FUZZY
       create the type definitions for range and relevance
   in the class constructor
           define the range, relevance, semantics
  if constructor is collection
   add a method to check exclusivity

  if constructor is alternative
   create the class name
   create the attributes of
   create additional boolean attributes for each component
   if attributes are of type multivalued, incompletevalued or  nullvalued
     inherit from class FUZZY
     create the type definitions for ranges and relevances
     in the class constructor
       define the ranges, relevances, semantics
       assign boolean attributes to false

  if constructor is sequence
   create the class name
   create the attribute of the class
   if attribute is of type multivalued, incompletevalued or nullvalued
    inherit from class FUZZY
    for the multivalued, incompletevalued or nullvalued attribute (if exists)
       create the type definitions for ranges and relevances
    in the class constructor
       define the ranges, relevances, semantics
```

The types supported by the ExIFO$_2$ conceptual model are also transformed into the FOOD structures for implementation. These types basically consist of *atomic*, *free* and *abstract* types. Since the atomic types correspond to the properties of the objects, their mappings are trivial. But for the *free* and *abstract* types, we show how there are mapped into the FOOD model is as follows:

```
  if type is free or abstract
   create the class name
   create the attributes for the components directed by the fragments
   if attributes are of type  multivalued, incompletevalued or  nullvalued
     inherit from class FUZZY
```

for the *multivalued, incomplete* or *nullvalued* attribute (if exists)
  create the type definition for ranges and relevances
in the class constructor
  define the ranges, relevances, semantics

Since ISA relationships are supported in the $ExIFO_2$ model, they also have to be checked and the necessary inheritance mechanism should incorporated into the FOOD model. Similarly, each fragment in the conceptual model has to be considered and mapped into the logical correspondences.

check for ISA relationships
  inherit from the class according to the type of ISA relationship
for the consistency of the fragments type
  add a method for checking consistency of fragments
for each of the *multivalued* attributes
  create the similarity matrix

Note that we create the similarity matrix for each multivalued attribute. If the multivalued attribute does not have fuzzy values, but only crisp values, then the identity matrix is assumed (note that the similarity relation is the generalization of the identity relation.)

Now we can map the conceptual model, shown in Figure 15, into a logical database model by applying this mapping algorithm. As we have mentioned before, the logical database model that we utilize here is the fuzzy object-oriented database (FOOD) model [5,14]. The resulting templates of the logical FOOD model after applying the mapping algorithm to the conceptual schema of the $EXIFO_2$ model for GIS applications is given in Appendix. We only give the result of the transformation without any further explanation about the resulting structures, since it is straightforward.

5 Conclusion

Nowadays, modeling needs for new applications and flaws in the currently used models have led to the definition of more powerful conceptual models. But, these models have still a number of serious limitations in implementing an increasing number of real-world applications that involve not only complex information (i.e., spatial), but also uncertainty and fuzziness. As a consequence, current research in this area is focusing on both handling fuzzy and complex information and defining new modeling and design approaches that are able to satisfy the needs for both traditional and advanced applications, such as GIS applications.

In this study, we mainly described a conceptual modeling approach for representing complex and uncertain geographic information by using an object-oriented paradigm. A conceptual schema specification for GIS applications is presented by utilizing the $ExIFO_2$ model. This model attempts to preserve the acquired strengths of semantic approaches, while integrating concepts of the object-oriented paradigm and fuzziness by including new constructors. Future work includes the design and development of a real geographic application by incorporating more geographic properties, continuous fields, and data input units and result presentation facilities into the model.

Acknowledgements

We thank to the anonymous reviewers of this paper for their helpful input.

References

1. Abiteboul, S. and R. Hull, "IFO: A Formal Semantic Database Model," ACM Trans. on Database Systems, Vol. 12, No.4, 1987, pp. 525-565.
2. Hadzilacos, T. and N. Tryfona, "An Extended Entity-Relationship Model for Geographic Applications," SIGMOD Record, Vol. 26, No.3, 1997, pp:24-29.
3. Vila, M.A., J.C. Cubero, J.M. Medina, and O. Pons, "A Conceptual Approach for Dealing with Imprecision and Uncertainty in Object-Based Data Models," International Journal of Intelligent Systems, Vol. 11, No.10, 1996, pp. 791-806.
4. Yang Li, Wu J, "Towards a Semantic Image Database System," Data & Knowledge Engineering, Vol. 22, 1997, pp:207-227.
5. Yazici, A. and R. George, *Fuzzy Database Modeling*, Physica-Verlag, Heidelberg New York, 1999.
6. Yazici, A and A. Cinar, "Conceptual Modeling for the Design of Fuzzy OO Databases" *Knowledge Management in Fuzzy Databases*, Edited by O. Pons, A. Vila and J. Kacprzyk, Physica-Verlag, Heidelberg.
7. Hammer, M. and D. McLeod, "Database Description with SDM: A Semantic Database Model,", ACM Transactions on Database Systems, Vol.6, No. 3, Sept. 1981, pp.351-386.
8. Shekhar, S., M. Coyle, B. Goyal, D. Liu, and S. Sarkar, "Data Models in Geographic Information Systems", Communication of ACM, Vol. 40, No.4, April 1997, pp: 103-111.
9. Shekhar, S., S.Chawla, S. Ravada, A. Fetterer, X. Liu, C. Lu " Spatial Databases – Accomplishments and Research Needs," IEEE Trans. on Knowledge and Data Engineering, Vol. 11, No.1, Jan/Feb 1999, pp:45-55.
10. Mohan L and Kashyap R.L., "An Object-Oriented Knowledge Representation for Spatial Information," IEEE Trans. on Software Engineering 14, No.5, 1988, pp:675-811.
11. Guting R.H., "An Introduction to Spatial Database Systems," Special Issue on Spatial Database Systems of the VLDB Journal, Vol. 3, No.4, Oct. 1994.
12. Gunther O., Riekert W.F., "The Design of GODOT: An Object-Oriented Geographic Information System," IEEE Data Engineering Bulletin 16, No.3, 1993.
13. Worboy, M.F., "Object-Oriented Approaches to Geo-Referenced Information," Int. J. Geographical Information System, Vol 8, No.4, 1994.
14. Yazici, George R., Aksoy D., "Extending the Similarity-Based Fuzzy Object-Oriented Data Model", Information Sciences (Int. Journal) Vol. 108, No.1-4, 1998, pp: 241-260.

15. Poncolet, P., M. Teissere, R. Cicchetti, and L. Lakhal, “Towards a Formal Approach for Object Database Design”, Proceedings of the 19th VLDB Conference, Dublin, Ireland, 1993, pp. 278- 289.
16. A. Motro, "Accommodating Imprecision in Database Systems: Issues and Solutions," SIGMOD RECORD, Vol. 19, No.4, 1990, pp. 69-74.
17. Zadeh, L.A., Similarity relations and fuzzy orderings. Information Sciences, Vol.3, No.2, 1971, pp:177-200.
18. Loucoupolous P, Zicari R., Conceptual Modeling, Databases and CASE: An Integrated View of Information Systems Development, Wiley Professional Computing, 1992.
19. Scholl M., Voisard A., Object-Oriented Database Systems for Geographic Applications: an Experiment with O_2, in: F. Bancilhon, C. Delobel and P. Kanellakis (Eds.), *The O_2 Book*, Morgan Kaufmann, San Mateo, Calif., 1992.

Appendix

This appendix represents the resulting class templates of the logical FOOD model after applying the mapping algorithm to the conceptual schema (shown in Figure 15) of the EXIFO$_2$ model for GIS applications.

```
typedef struct {
    AnsiString        rangecolor;
    AnsiString        rangetexture;
} RangesGraphProp;

typedef struct {
    float     RLVcolor;
    float     RLVtexture;
} RLVGraphProp;

typedef struct {
    incompletevalued <int>RangeIncompval;
}RangesIntVal;

typedef struct {
    float RangeIncompval;
}RLVIntVal;

AnsiString  colordomain[] = { }
AnsiString  texturedomain[] = { }
Float colormatrix [][]= { }
Float texturematrix [] []= { }

Class Tmap {
    Public:
            Char        *name;
            Float       scale;
            Tobjects    *Gobjects;
    TMap();
}

Class TGobjects: SET {
    Public:
            TGobject     *Gobject;
    TGobjects () {
            Gobject= new TGobject;
    }
    booelan check_collection_Gobject();
}
```

```
Class TGobject {
   Public:
            TGraphProp      *Graphical;
            TDescription    *Description;
            TGeos           *Geos;
   TGobject();
}

Class TGraphProp: FUZZY {
   Public:
            Int     Font;
            Int     lnwdth;
            Char    *text;
            Hist    histrogram;
            Multivalued <ansistring> color;
            Multivalued <ansistring> texture;
            RangesGraphProp         Ranges;
            RLVGraphProp  Relevance;

TGraphProp() {
            Ranges.rangecolor ="dark blue, blue, dark green, green";
            Ranges.rangetexture="rectangles, circular lines";
            Relevance.RLVcolor=0.1;
            Relevance.RLVtexture=0.1;
            Color.semantics="OR";
            Texture.semantics="OR";
   }
}

Class TDescription {
   Public:
            TAspatials      *Aspatials;
   TDescription();
}

Class TAspatials: SET {
   Public:
            TAspatial       *Aspatial;
   TAspatials() {
            Aspatial=new TAspatial;
   }
}

Class TAspatial {
   Public:
            Char *name;
            TAspatial();
}
```

```
Class TIntAttr {
   Public:
            Int       val;
   TIntAttr();
}

Class TStrAttr {
   Public:
            Char   *val;
   TStrAttr();
}

Class TIncomAttr: FUZZY {
   Public:
            Boolean is_crispval;
            Boolean is_incompval;
            Incomletevalued<int> incompval;
            RangeIntVal       Ranges;
            RLVIntVal         Relevance;

TIncomAttr() {
            Ranges.incompval=    ;
            Relevance.incompval=  ;
            Is_crispval=false;
            Is_incompval=false;
   }
}

Class  TGeos {
   Public:
            TGeo     *Geo;
   TGeos();
}

Class TGeo { }

Class TRegion : TGeo {
   Public:
            Int       regid;
            Tline     *line;
   TRegion();
   Private:
   Boolean is_complex (Tregion, TLine);
}

Class TRegWithHoles: TRegion {
   Public:
            Int       numofholes;
   TRegionWithHoles();
}
```

```
Class TLine: Geo {
   Public:
            Int        lineid;
            TBegEnd          *BeginEnd;
            TRegion          *region;
   TLine();
   Private:
            Boolean  is_partcomp(Tline, Tregion);
}

Class TBegEnd {
   Public:
            TBegin     *begin;
            TEnd       *end;
   TBegEnd();
}

Class TBegin: Point { }

Class TEnd: Point {}

Class TPoint: Geo {
   Public:
            Int          pointid;
            TCoord   *Coord;
   TPoint();
}

Class TCoord {
   Public:
            Int  x;
            Int  y;
   TCoord();
}
```

Part IV.

Fuzzy Database Models: Definition and Implementation

A Generalised Object-Oriented Database Model

Guy de Tré, Rita de Caluwe, and Bart Van der Cruyssen

University of Ghent, Computer Science Laboratory, Sint-Pietersnieuwstraat 41, B-9000 Ghent, Belgium

Summary. A new formal framework for a generalised object-oriented database model that is able to cope with fuzzy and uncertain information is presented. The generalised model is obtained as a generalisation of a crisp object-oriented database model that is compliant with the ODMG de facto standard and is built upon an algebraic type system and a constraint system. The novelty of our approach is the integration of (generalised) constraints to enforce integrity rules and to specify the formal semantics of the database model.

1 Introduction

During the past decade several "fuzzy" object-oriented database models have been proposed. An overview of them can be found in [1]. These models do not conform to a single underlying object data model, which is a consequence of the present lack of object standards.

As far as crisp object-oriented databases are concerned, the ODMG proposal [2] (which tends to become the de facto object data model standard) offers new promising perspectives. However, it still suffers from some shortcomings such as the absence of formal semantics [3] and its limited ability to deal with constraints, despite the fact that a thorough support of constraints is the most obvious way to guarantee the integrity of a database.

The formal framework we present is ODMG compliant and moreover overcomes the mentioned ODMG shortcomings; it also supports the management of both fuzzy and uncertain information. In our approach, we first build a crisp formal type system and constraint system which allows us to give formal object and database scheme definitions. Next, the type system and the constraint system are generalised using the powerful notion of generalised constraints as they have been defined by L.A. Zadeh [4]. This generalisation allows us to define generalised object and database schemes, which are the main components of our generalised object-oriented database model. All the generalised concepts are defined in accordance with the directions of the UFO model [5].

Here, we focus on the definition of the generalised data model. A number of examples drawn from the medical field are included to show the practical interest of the model.

2 A Basic Formal Framework for Object-Oriented Databases

Over the last years, several formal models for the concept of crisp object-databases have been published. Among them we mention the approach based on Method Schema's [6], the Reloop Algebra [7], the Association Algebra [8], the F-logic [9], the Monoid Calculus [10], and the CROQUE-model [11]. However, none of these integrate all the concepts of the ODMG proposal, neither do they deal with constraints. Some of them have been extended towards a fuzzy database model [8,9]. Nevertheless, since crisp databases can be seen as special cases of fuzzy and/or uncertain databases, our aim is to define a generalisation of a crisp object-oriented database model (rather than defining an extension of a crisp database model), according to the principles of the UFO model [5].

In the following subsections the main components of a basic formal framework are described. First, we introduce the type system and its instances (Subsection 2.1). Next, we give the definition of the constraint system and its instances (Subsection 2.2) and finally, we define object schemes and database schemes (Subsection 2.3).

2.1 Types and Type System

To support the definition of types, a type system is built. The definition of the type system we propose is similar to the one given in [12] but completely compliant with the ODMG specifications. Starting with the specification of base types, we build a type system, by stating rules which define the specification of object types, generic collection literal types and construction literal types. Additionally, we specify reference types and the Void type. The defined type system is object-oriented, supporting both multiple inheritance and inclusive polymorphism. An overview of our type system is given in Figure 1.

The most primitive type of the system is the so-called Void type. This type is introduced for the compliancy with the ODMG de facto standard and will be used in situations where no other types are applicable. Next, reference types which enable to refer to the instances of object types are defined. A distinction between single-valued and multi-valued reference types is made, based on whether the type allows to refer to one or more instances of an object type. Reference types are used to formalize the relationships between the object types in the database scheme. A literal type is defined as either a base type (boolean, integer, octet, real or string), a collection literal type (set, bag, list, array or dictionary) or a construction literal type (enumeration type or structured literal type). Finally, we have the definition of the object types, which can be seen as the most advanced types of the type system.

Types. Each type in our formal framework is defined by a type specification and one or more type implementations. We do not limit ourselves to

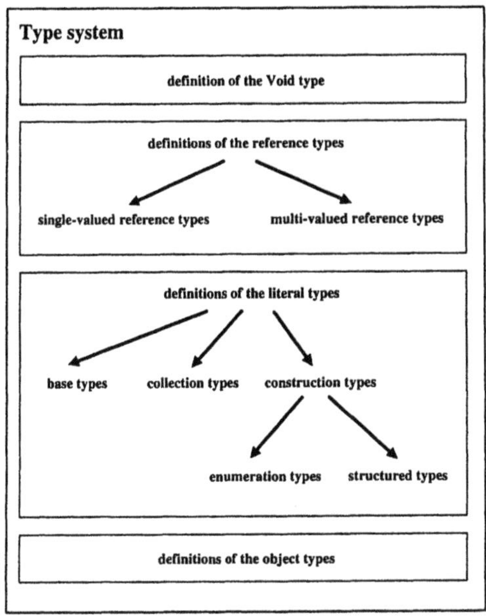

Fig. 1. The type system: an overview

the syntax of the type specifications, but also endow the definition rules with formal semantics.

- *Syntax of a type (specification):* The syntax of the Void type is given by the symbol **Void**. The set of expressions that satisfy the syntax of a reference type are denoted as $T_{reference}$, the set of expressions that satisfy the syntax of a literal type as $T_{literal}$, and the set of expressions that satisfy the syntax of an object type as T_{object}.
- *Semantics of a type (specification):* Our semantic definitions are related to domains and sets of domains (cf. [13]). More specifically, the formal semantics of a type specification ts are determined by:
  - a set of domains D_{ts},
  - a designated domain $dom_{ts} \in D_{ts}$,
  - a set of operators O_{ts},
  - and a set of axioms A_{ts}.

  The designated domain dom_{ts} is called the domain of the type (specification) ts. Each domain dom_{ts} consists of the set of all the possible values of the type (specification) ts. In our basic formal framework, a domain consists of crisp values and contains a bottom value $\perp_{ts}$ which represents the "undefined" value. The set of operators O_{ts} contains all the operators that are defined on the domain dom_{ts}, the set of domains D_{ts} consists of all the domains that are involved in the definition of the operators of O_{ts}, whereas the set of axioms A_{ts} contains all the axioms that are involved in the definition of the semantics of O_{ts}.

Example 1. As an example, the definition of the specification of the base type integer is given:

- syntax:

  Integer

- semantics:
  1. the designated domain:

  $$dom_{Integer} = \mathbb{Z} \cup \{\perp_{Integer}\}$$

  2. the set of domains $D_{Integer}$:

  $$D_{Integer} = \{dom_{Boolean}, dom_{Integer}\}$$

  3. the set of operators $O_{Integer}$:

  $$\begin{aligned} O_{Integer} = \{ &= : (dom_{Integer})^2 \rightarrow dom_{Boolean}, \\ &\neq : (dom_{Integer})^2 \rightarrow dom_{Boolean}, \\ &< : (dom_{Integer})^2 \rightarrow dom_{Boolean}, \\ &> : (dom_{Integer})^2 \rightarrow dom_{Boolean}, \\ &\leq : (dom_{Integer})^2 \rightarrow dom_{Boolean}, \\ &\geq : (dom_{Integer})^2 \rightarrow dom_{Boolean}, \\ &+ : (dom_{Integer})^2 \rightarrow dom_{Integer}, \\ &- : (dom_{Integer})^2 \rightarrow dom_{Integer}, \\ &* : (dom_{Integer})^2 \rightarrow dom_{Integer}, \\ &\mathbf{div} : (dom_{Integer})^2 \rightarrow dom_{Integer}, \\ &\mathbf{mod} : (dom_{Integer})^2 \rightarrow dom_{Integer}, \\ &\perp : () \rightarrow dom_{Integer}\} \end{aligned}$$

  4. the axioms of $A_{Integer}$:
     (a) the semantics of the binary operators $=$, $\neq$, $<$, $>$, $\leq$ and $\geq$

     Because $dom_{Integer} \setminus \{\perp_{Integer}\}$ equals the totally ordered set of integer numbers $\mathbb{Z}$, it is stated that the binary operators $=$, $\neq$, $<$, $>$, $\leq$ and $\geq$ have the same semantics as their counterparts $=$, $\neq$, $<$, $>$, $\leq$ and $\geq$ within $\mathbb{Z} \times \mathbb{Z}$ when restricted to the set:

     $$dom_{Integer} \setminus \{\perp_{Integer}\}$$

For undefined argument values the semantics become:

$$\forall\, x \in dom_{Integer} : \quad relat(x, \perp_{Integer}) = relat(\perp_{Integer}, x) = \perp_{Boolean}$$

where *relat* is a variable copula whose values are respectively the symbols $=$, $\neq$, $<$, $>$, $\leq$ and $\geq$;

(b) the semantics of the binary operators $+$, $-$, $*$, **div** and **mod**

For the same reason as above, it is stated that the operators $+$, $-$, $*$, **div** and **mod** have the same semantics as their counterparts within $\mathbb{Z} \times \mathbb{Z}$ when restricted to the set:

$$dom_{Integer} \setminus \{\perp_{Integer}\}$$

For undefined argument values the semantics become:

$$\forall\, x \in dom_{Integer} : \quad op(x, \perp_{Integer}) = op(\perp_{Integer}, x) = \perp_{Integer}$$

where *op* is a variable copula whose values are respectively the symbols $+$, $-$, $*$, **div** and **mod**;

(c) the truth table for the null-ary operator $\perp$ is

$\perp$	
	$\perp_{Integer}$

the bottom operation results in an undefined integer value. ◇

The type system. The type system TS, which allows to derive both the specification and the implementations of the valid types, is formally defined as a triple:

$$TS = [T, P, f_{impl}^{type}]$$

where $T = \{\mathbf{Void}\} \cup T_{reference} \cup T_{literal} \cup T_{object}$ is the set of valid type specifications, P is the infinite set of all programs —indepedent of the language in which they are written— and f_{impl}^{type} is called the (type) implementation function. This function is defined as a mapping

$$\begin{aligned} f_{impl}^{type} : T &\rightarrow \wp(P) \\ ts &\mapsto \{p_1, \ldots, p_n\} \end{aligned}$$

which maps each type specification ts of the domain T onto the subset $\{p_1, \ldots, p_n\}$ which is an element of the co-domain $\wp(P)$ (the powerset of P) that contains all the implementations of ts.

Instances of the type system. An instance t of the type system TS is called a type and is formally defined as a couple:

$$t = [ts, f_{impl}^{type}(ts)], \text{ with } ts \in T$$

t is called a void type when $ts =$ **Void**; t is called a literal type when $ts \in T_{literal}$; t is called a reference when $ts \in T_{reference}$; otherwise t is called an object type (cf. Figure 2).

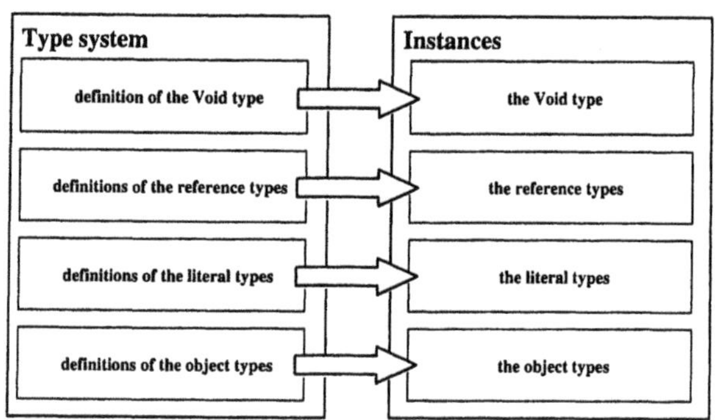

Fig. 2. The instances of the type system

Here we focus on the object types, as they provide a formal basis for the most important notion in ODMG, namely the object. However, as well as the properties, the behaviour of both objects and literals is formalised by the types of our type system. Obviously, the behaviour of literals is restricted to the implicitly defined operators.

The syntax of an arbitrary object type $[ts, f_{impl}^{type}(ts)], ts \in T_{object}$ is defined as either:

$$\textbf{Class } id(id_1 : s_1; \ldots ; id_n : s_n)$$

or

$$\textbf{Class } id : \widehat{id}_1, \ldots, \widehat{id}_m(id_1 : s_1; \ldots ; id_n : s_n)$$

In these definitions, id is an identifier that represents the name of the object type, n and m are integer values, $\{\widehat{id}_i | i = 1, 2, \ldots, m\}$ is the finite set of identifiers representing the parent types of the object type and $\{id_i : s_i | i = 1, 2, \ldots, n\}$ is the finite set of the characteristics of the object type which are explicitly specified within the syntax (beside these characteristics, the object type can inherit other characteristics from its parent types).

For each characteristic $id_i : s_i, i = 1, 2, \ldots, n$, id_i is called the identifier of the characteristic and s_i is the specification of the characteristic. This specification can either be:

- a literal type (i.e. $s_i \in T_{literal}$) in which case the characteristic is called an attribute;
- a reference type (i.e. $s_i \in T_{reference}$) in which case the characteristic is called a reference;
- a signature in which case the characteristic is called a method.

Example 2. With

$$\{\mathbf{String}, \mathbf{Real}, \mathbf{List}(ts), \mathbf{Set}(ts), \mathbf{Enum}\ TSex(male = 0, female = 1), \\ TDate, \mathbf{Struct}\ TResult(Value : \mathbf{Real}; Date : TDate)\} \subset T_{literal}$$

and

$$\{\mathbf{Set}(\mathbf{Ref}(ts)), \mathbf{List}(\mathbf{Ref}(ts))\} \subset T_{reference}$$

the specifications of the object types

$$[TMedication, f^{type}_{impl}(TMedication)], [TPerson, f^{type}_{impl}(TPerson)] \\ \text{and } [TPatient, f^{type}_{impl}(TPatient)]$$

of a simplified CPR (Computer based Patient Record) can be defined as:

$$\begin{aligned} \mathbf{Class}\ TMedication(&Name : \mathbf{String}; \\ &Composition : \mathbf{Set}(\mathbf{String}); \\ &Users : \mathbf{Set}(\mathbf{Ref}(TPatient))) \end{aligned}$$

$$\begin{aligned} \mathbf{Class}\ TPerson(&Name : \mathbf{String}; \\ &Given\_names : \mathbf{List}(\mathbf{String}); \\ &Birth\_date : TDate; \\ &Sex : TSex; \\ &Parents : \mathbf{Set}(\mathbf{Ref}(TPerson)); \\ &Children : \mathbf{List}(\mathbf{Ref}(TPerson)); \\ &Add\_child : \\ &\quad \mathbf{Signat}((New\_child : \mathbf{Ref}(TPerson)) \rightarrow \mathbf{Void})) \end{aligned}$$

and

$$\begin{aligned}
&\textbf{Class } TPatient : TPerson(\\
&\qquad Height : \textbf{List}(TResult); \\
&\qquad Weight : \textbf{List}(TResult); \\
&\qquad Allergies : \textbf{Set}(\textbf{String}); \\
&\qquad Medical_history : \textbf{List}(\textbf{String}); \\
&\qquad Cur_presc : \textbf{Set}(\textbf{Ref}(TMedication)); \\
&\qquad Process_results : \\
&\qquad\quad \textbf{Signat}((Results : \textbf{List}(TResult)) \rightarrow \textbf{Void}); \\
&\qquad Add_child : \\
&\qquad\quad \textbf{Signat}((New_child : \textbf{Ref}(TPatient)) \rightarrow \textbf{Void})) \quad \diamond
\end{aligned}$$

Instances of a type. As illustrated in Figure 3, the instances of a reference type, a literal type and an object type are respectively called reference instances, literals and objects, whereas the Void type cannot have instances at all. Hereafter, only instances of object types will be handled in detail.

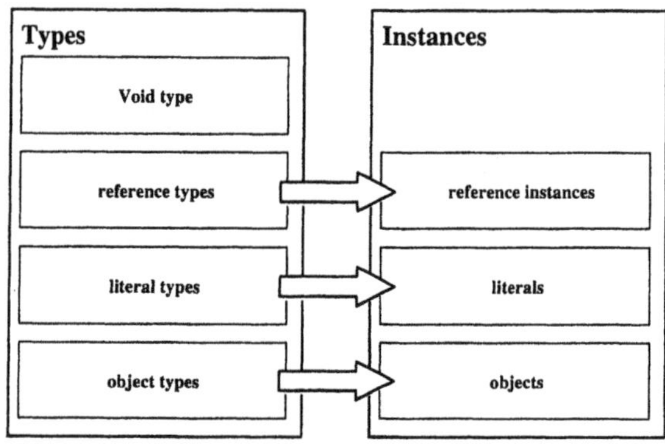

Fig. 3. The instances of a type

Depending on its lifetime, an object can either be transient or persistent. A transient object o is defined as a pair

$$o = [t, v] \text{ with } t = [ts, f_{impl}^{type}(ts)], ts \in T_{object}$$

whereas a persistent object is defined as a quadruple

$$o = [oid, N, t, v] \text{ with } t = [ts, f_{impl}^{type}(ts)], ts \in T_{object}$$

in these definitions:

- t is the type of the object,
- $v \in dom_{ts}$ is the state of the object,
- oid is a unique object identifier and
- N is a set of object names.

For every attribute or relationship $id_i : s_i$ specified within the type specification ts, the state v contains a value $v_i \in dom_{s_i}$ and the set of object names N can be empty.

The extent V_t^{extent} of an object type t is the set of all the persistent objects of t within a particular database. If type t is a subtype of type $\widehat{t}$, then the extent of t is a subset of the extent of $\widehat{t}$.

Example 3.

$$
\begin{aligned}
[oid1, \{P721028\text{–}1101\}, TPatient, \mathbf{Struct}(&\\
Name :& \text{"Smith"};\\
Given\_names :& \mathbf{List}(\text{"Ann"});\\
Birth\_date :& 28.10.1972;\\
Sex :& female;\\
Parents :& \bot_{Set(Ref(TPerson))};\\
Children :& \mathbf{List}(oid2, oid3);\\
Height :& \mathbf{List}(\mathbf{Struct}(Value : 1.72; Date : 06.04.1999));\\
Weigth :& \mathbf{List}(\mathbf{Struct}(Value : 63; Date : 06.04.1999),\\
& \qquad\qquad \mathbf{Struct}(Value : 60; Date : 21.06.1999));\\
Allergies :& \mathbf{Set}(\text{"penicillin"}, \text{"cats"});\\
Medical\_history :& \mathbf{List}(\text{"influenza on 02.02.1998"});\\
Cur\_presc :& \mathbf{Set}(\,))]
\end{aligned}
$$

is the instance of the object type $TPatient$ for the patient "Ann Smith" identified internally by the object identifier $oid1$ and externally by the object name $P721028 - 1101$. ◇

2.2 Constraints and Constraint System

Crisp constraints can be seen as Boolean functions and will be used to specify the formal semantics of a database. Typically, they indicate if a given object has a valid state within the context of a given database (e.g. it satisfies all business rules). A constraint is formally defined as an instance of a constraint system.

Different kinds of constraints are distinguished. A first distinction is based on whether their definition involves other related object types or not (single-type dependent or multi-type dependent). A second distinction is based on whether only a single instance or the entire extent of an object type is involved in the evaluation of the constraint. This results in four sets of constraint specifications: $C^{single}_{instance}$, C^{single}_{extent}, $C^{multi}_{instance}$ and C^{multi}_{extent} (cf. Figure 4).

CONSTRAINT SYSTEM $\xrightarrow{\text{instance}}$ CONSTRAINT

. Single-type dependent constraint for a single instance
  . Domain constraint
  . Not null constraint
  . Single-type dependent value constraint
  . Single-type dependent transition constraint
. Singe-type dependent constraint for the extent
  . Key constraint
  . Single-type dependent aggregate constaint
. Multi-type dependent constraint for a single instance
  . Multi-type dependent value constraint
  . Multi-type dependent transition constraint
. Multi-type dependent constraint for the extent
  . Unicity constraint
  . Referential constraint
  . Multi-type dependent aggregate constraint

Fig. 4. The constraint system: an overview

Constraints. Each constraint in our formal framework is defined by a constraint specification and one or more constraint type implementations.

Within the set $C^{single}_{instance}$ we distinguish:

- *Domain constraints* define the set of all possible values of an object type. As a consequence, they also limit the possible values for the attributes, the references and the components of the structured attributes of the object type. Since domain constraints match the definition of the designated domain within the definition of the semantics of the object type, they are not (re)defined as elements of $C^{single}_{instance}$.
- *Not null constraints* are also defined for an attribute or a relationship of an object type and evaluate to "true" when the value of the characteristic (within the state of the object) differs from the "undefined" value $\perp_{ts}$ of the domain of the type of the characteristic; otherwise they evaluate to "false" (e.g., the null constraint

  $$\rho^{not\_null}_{\{TPerson.Name\}}[\,]$$

  enforces that the values for the attribute $Name$: **String** differ from $\perp_{String}$).
- *Single-type dependent value constraints* are defined for an attribute of an object type. Their definition contains a boolean expression which is defined over the attributes of the object type and can evaluate to "true", "false" or $\perp_{Boolean}$. The constraint evaluates to "true" for an object when

the expression evaluates to "true" taking the values of the state of the object as arguments (e.g., the single-type dependent value constraint

$$\rho^{value}_{\{TPatient.Height.Value\}}[0 < TPatient.Height.Value < 2.6]$$

enforces that the value for the height of a patient must be less than 2.6 (metre)).

- *Single-type dependent transition constraints* are similar to the single-type dependent value constraints, except that their expression always defines a condition for the transition between the old and new values of the attribute (e.g., the single-type dependent transition constraint

$$\rho^{trans}_{\{TPatient.Allergies\}}[(TPatient.Allergies)^{new} \supseteq TPatient.Allergies]$$

can be used to enforce that the previous set of allergies of a patient always must be a subset of the new one).

Within the set C^{single}_{extent}, we distinguish:

- *Key constraints* are used to define a key —i.e. an attribute, relationship or set of attributes and/or relationships of an object type whose value(s) might be used to uniquely identify the instances of the object type within the scope of its extent— (e.g., with the key constraint

$$\rho^{key}_{\{TPerson\}}[Name, Given\_names]$$

each combination of "Name"- and "Given_names"- values must be unique within the extent of the *TPerson* type).

- *Single-type dependent aggregate constraints* are defined for the extent of an object type. Their definition contains a boolean expression that enforces a condition on the extent. Again, the constraint can evaluate to "true", "false" or $\perp_{Boolean}$ depending on the evaluation of the expression (e.g., the constraint

$$\rho^{aggr}_{\{TPerson\}}[\mathbf{count}(V^{extent}_{TPerson}) \leq 1000]$$

enforces that the extent $V^{extent}_{TPerson}$ of the object type *TPerson* cannot have more than thousand instances).

Within the set $C^{multi}_{instance}$, one can find:

- *Multi-type dependent value constraints* are the counterparts of the single-type dependent value constraints. They are also defined for an attribute of an object type, but their expression is defined over at least one attribute of a related object type (e.g., the multi-type dependent value constraint

$$\rho^{value}_{\{TPatient.Cur\_presc.Composition,\{TPerson,TPatient,TMedication\}\}}[\mathbf{is\_empty}(\cap(TPatient.Cur\_presc.Composition, TPatient.Allergies))]$$

can be used to prevent the prescription of medication the patient is allergic to).

- *Multi-type dependent transition constraints* are the counterparts of the single-type dependent transition constraints, which means that their expression is defined over at least one attribute of a related object type.

Within the set C^{multi}_{extent}, we distinguish:

- *Uniqueness constraints* are used to guarantee the uniqueness of the object identifier (*oid*) and the object names ($\in N$) of an object within the scope of the union of the extents of a set of object types (e.g., with the uniqueness constraint

$$\rho^{name}_{\{TPerson,\{TPerson,TPatient,TMedication\}\}}[\,]$$

the object names of $TPerson$ objects must be unique within the union of the extents of the object types $TPerson$, $TPatient$ and $TMedication$).
- *Referential constraints* are defined to maintain the referential integrity of the (binary) relationships between objects by guaranteeing that an object refers to existing objects only (e.g., the constraint

$$\rho^{reference}_{\{\{TPatient.Cur\_presc,TMedication.Users\},\{TPatient,TMedication\}\}}[\,]$$

is used to enforce the referential integrity of the Patient-Medication relationship).
- *Multi-type dependent aggregate constraints* are the counterparts of the single-type dependent aggregate constraints. They are also defined for the extent of an object type, but their expression is defined over at least one attribute of a related object type or the extent of at least one other object type.

The constraint system. The constraint system CS, which allows to derive both the specification and the implementations of the constraints, is defined as a triple:

$$CS = [C, P, f^{constr}_{impl}]$$

where $C = C^{single}_{instance} \cup C^{single}_{extent} \cup C^{multi}_{instance} \cup C^{multi}_{extent}$ is the set of valid constraint specifications, and P and f^{constr}_{impl} denote respectively the infinite set of all programs and the (constraint) implementation function, which is defined as:

$$\begin{aligned} f^{constr}_{impl} &: C \rightarrow \wp(P) \\ & cs \mapsto \{p_1, \ldots, p_n\} \end{aligned}$$

Hereby, f^{constr}_{impl} is a mapping which maps each constraint specification cs of the domain C onto the subset $\{p_1, \ldots, p_n\}$ of the co-domain $\wp(P)$ (the powerset of P) that contains all the implementations of cs.

Instances of the constraint system. An instance c of the constraint system CS is called a constraint and is defined as a pair:

$$c = [cs, f_{impl}^{constr}(cs)], \text{ with } cs \in C$$

2.3 Object Schemes and Database Schemes

The structure of the adopted scheme definitions is similar to the one for relational databases presented in [14]. Both the definitions of object scheme and database scheme rely on the definitions of types and constraints.

An object scheme and its instances. The full semantics of an object are described by an object scheme os. A scheme completely defines the object, including the constraints that apply to the object. An object scheme is formally defined as a triple:

$$os = [t, M, C_t]$$

in which

- t is an object type,
- M represents the meaning of the object scheme and
- C_t is a finite set of constraints.

The meaning M is an "informal component of the definition, since it refers to the real world and since it will be described in a natural language" [14]. (Although a human comprehensible description has little theoretical importance, it is included as a fundamental time independent property of the object scheme.) The set of constraints C_t is defined as a finite subset of $C_{instance}^{single}$. Every constraint $c \in C_t$ must be defined for a single instance of t.

An instance of the object type t is defined to be an instance of the object scheme $os = [t, M, C_t]$ if and only if it satisfies all the constraints of C_t and all the constraints of the sets $C_{\hat{t}}$ of the object schemes $[\hat{t}, \widehat{M}, C_{\hat{t}}]$ which have been defined for the supertypes $\hat{t}$ of t.

Example 4. With respect to the object types

$$[TPerson, f_{impl}^{type}(TPerson)] \text{ and } [TPatient, f_{impl}^{type}(TPatient)]$$

of Example 2, the object schemes $OSPerson$ and $OSPatient$ can be constructed as:

$$OSPerson = [TPerson, M_{OSPerson}, C_{TPerson}]$$

in which:

- $TPerson$ is the type of the object scheme;

- the meaning $M_{OSPerson}$ is given as:

  "This object scheme is used as the specification of an abstraction for persons, such as they appear in the real world.";
- and the set $C_{TPerson}$ is constructed as:

$$C_{TPerson} = \{$$
$$\rho^{not\_null}_{\{TPerson.Name\}}[\,],$$
$$\rho^{value}_{\{TPerson.Birth\_date\}}[01.01.1860 \leq TPerson.Birth\_date \leq Today]\}$$

and

$$OSPatient = [TPatient, M_{OSPatient}, C_{TPatient}]$$

in which:

- $TPatient$ is the type of the object scheme;
- the meaning $M_{OSPatient}$ is given as:

  "This object scheme is used as the specification of a simplified abstraction of CPR's as they exist in the cabinet of a physician."
- and the set $C_{TPatient}$ is constructed as:

$$C_{TPatient} = \{$$
$$\rho^{value}_{\{TPatient.Height.Value\}}[0 < TPatient.Height.Value < 2.6],$$
$$\rho^{value}_{\{TPatient.Weight.Value\}}[0 < TPatient.Weight.Value < 500],$$
$$\rho^{trans}_{\{TPatient.Allergies\}}[$$
$$(TPatient.Allergies)^{new} \supseteq TPatient.Allergies]\} \quad \diamond$$

A database scheme and its instances. A database scheme ds describes the information which is stored in the database and is defined as a triple:

$$ds = [D, M, C_D]$$

in which

- $D = \{os_i | i = 1, 2, \ldots, n, n \in \mathbb{N}_0\}$ is a finite set of object schemes, i.e. $os_i = [t_i, M_i, C_{t_i}], i = 1, 2, \ldots, n,$
- M represents the meaning of ds and
- C_D is a finite set of constraints which associates extra conditions with the instances of the set of object schemes.

Every object scheme in D is defined for a different object type. If an object scheme $os_i \in D$ is defined for an object type t and t' is a supertype of t or if there exists a reference between t' and t, then there exists an object scheme $os'_i \in D$ which is defined for t'. The meaning M will be described in a natural language and can therefore be seen as an informal component. The set of constraints C_D is a subset of $C^{single}_{extent} \cup C^{multi}_{instance} \cup C^{multi}_{extent}$. For every object scheme $os_i \in D$, there exist uniqueness constraints in C_D which guarantee the uniqueness of the object identifiers and object names of the instances of os_i. Finally, every constraint $c \in C_D$ is defined for the instances (or the extent) of the object type t_i of an object scheme $os_i \in D$.

An instance of an object scheme $os_i \in D$ is defined to be an instance of the database scheme $ds = [D, M, C_D]$ if and only if it satisfies all the constraints of C_D. The extent of a database scheme ds is called a database and is defined as the set of sets of database instances obtained by grouping all instances per object scheme.

Example 5. With respect to the object schemes $OSPerson$ and $OSPatient$ of Example 4 and an object scheme $OSMedication$ —which is derived in a similar way from the object type $TMedication$—, the database scheme $DSCpr$ can be built as:

$$DSCpr = [DCpr, M, C_{DCpr}]$$

in which:

- $DCpr = \{OSPerson, OSPatient, OSMedication\}$ is the finite set of object schemes;
- the meaning M is given as:

  "This database scheme is used as a specification of a simplified computer based patient record system (CPR)."
- and the set C_{DCpr} is constructed, with the shorthand notation

  $$T_{DCpr} = \{TPerson, TPatient, Medication\}$$

  as:

  $$\begin{aligned} C_{DCpr} = \{ & \rho^{oid}_{\{TPerson, T_{DCpr}\}}[\,], \\ & \rho^{name}_{\{TPerson, T_{DCpr}\}}[\,], \\ & \rho^{oid}_{\{TPatient, T_{DCpr}\}}[\,], \\ & \rho^{name}_{\{TPatient, T_{DCpr}\}}[\,], \\ & \rho^{oid}_{\{TMedication, T_{DCpr}\}}[\,], \\ & \rho^{name}_{\{TMedication, T_{DCpr}\}}[\,], \\ & \rho^{reference}_{\{\{TPerson.Parents, TPerson.Children\}, T_{DCpr}\}}[\,], \\ & \rho^{reference}_{\{\{TPatient.Cur\_presc, TMedication.Users\}, T_{DCpr}\}}[\,], \\ & \rho^{key}_{\{TPerson\}}[Name, Given\_names], \\ & \rho^{key}_{\{TPatient\}}[PatientID]\} \end{aligned}$$ ◇

In the previous example, two bidirectional referential constraints are defined: one for the pair of references:

$$(Parents : \textbf{Set}(\textbf{Ref}(TPerson)), Children : \textbf{List}(\textbf{Ref}(TPerson)))$$

of the object type $TPerson$ and one for the pair of references

$$(Cur\_presc : \textbf{Set}(\textbf{Ref}(TMedication)), Users : \textbf{Set}(\textbf{Ref}(TPatient)))$$

of the object types $TPatient$ and $TMedication$. By doing this, a pair of references is "promoted" to a single relationship, conform to the ODMG report.

3 A Formal Framework for Generalised Object-Oriented Databases

By analogy with the fact that a crisp database can be seen as a special case of a fuzzy and/or uncertain database, our generalised database model is constructed as a generalisation of a crisp one.

The basic formal framework of Subsection 2 has been choosen as a starting point. First, the type system has been generalised by generalising the type definitions. Secondly, the constraint system has been generalised. The generalised type system provides (among others) the definition of a generalised object type, whereas the generalised constraint system provides the generalisation of constraints. Both of them take part in the definition of the concepts of generalised object scheme and of generalised database scheme (cf. Figure 5).

3.1 Generalised Types and Generalised Type System

A generalised (fuzzy) type system is obtained by generalising the definition of the crisp type system $TS = [T, P, f_{impl}^{type}]$. Hereby, the set of type specifications T is generalised to a set $\tilde{\tilde{T}}$ by generalising the definitions of the domains and operators of the type specifications $ts \in T$.

Generalised types. For every type specification $ts \in T$, a generalised counterpart $\tilde{\tilde{ts}} \in \tilde{\tilde{T}}$ is defined by generalising the definitions of its syntax and its semantics.

- *Syntax of a generalised type (specification):* The syntax of the generalised Void type is given by the symbol $\tilde{\tilde{\textbf{Void}}}$. The set of all the expressions that satisfy the syntax of a generalised reference type are denoted as $\tilde{\tilde{T}}_{reference}$, the set of all the expressions that satisfy the syntax of a

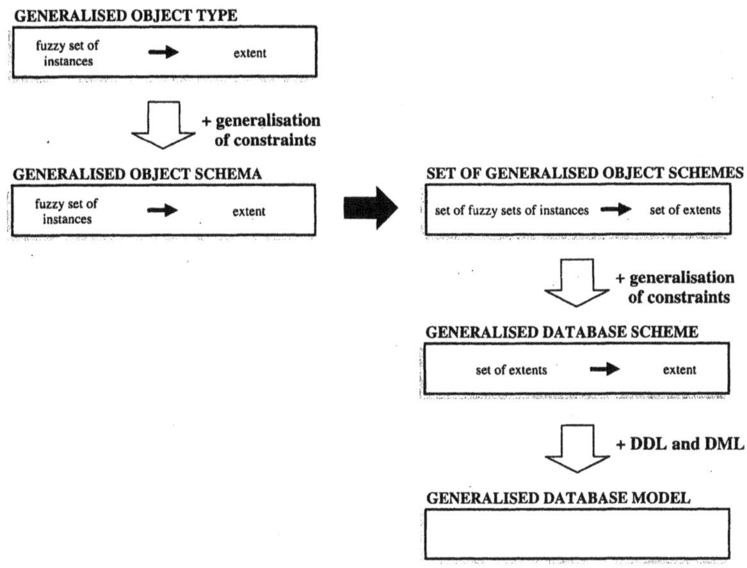

Fig. 5. Our generalised formal framework: an overview

generalised literal type as $\tilde{\tilde{T}}_{literal}$, and the set of all the expressions that satisfy the syntax of a generalised object type as $\tilde{\tilde{T}}_{object}$.
Furthermore, the definitions of the syntax $id : ts$ of an attribute or a reference of an object type, and the syntax $id : ts$ of a component of a structured literal type are generalised through the use of generalised constraints as defined by L.A. Zadeh [4]: the syntax of a generalised attribute, reference or component is defined as

$$id \; isr \; \tilde{\tilde{ts}}$$

in which isr is a variable copula and r is a discrete variable whose value defines the way the values of the (domain of the) characteristic or component are constrained. For now, the considered values of r are e, *blank* and v. The semantics of these notations will be clarified below.

- *Semantics of a generalised type (specification):* The formal semantics of the generalisation $\tilde{\tilde{ts}}$ of a type (specification) ts are determined by:
  - a set of domains $D_{\tilde{\tilde{ts}}}$,
  - a designated domain $dom_{\tilde{\tilde{ts}}} \in D_{\tilde{\tilde{ts}}}$,
  - a set of operators $O_{\tilde{\tilde{ts}}}$,
  - and a set of axioms $A_{\tilde{\tilde{ts}}}$.

  The designated domain $dom_{\tilde{\tilde{ts}}}$ of a generalised type $\tilde{\tilde{ts}}$ has been defined as an extension of the designated domain dom_{ts} of its crisp counterpart ts.

For the Void type, the reference types and the literal types, $dom_{\tilde{\tilde{ts}}}$ is defined as the generalisation

$$dom_{\tilde{\tilde{ts}}} = \tilde{\wp}(\tilde{\wp}(dom_{ts}))$$

and thus consists of fuzzy sets $\tilde{\tilde{V}}_i$ which have been defined on the set of all the fuzzy sets $\tilde{V}_j$ which are defined on the crisp domain dom_{ts}. The membership degrees of the elements of the fuzzy sets $\tilde{\tilde{V}}_i$ are interpreted as degrees of uncertainty, whereas the membership degrees of the elements of the fuzzy sets $\tilde{V}_j$ are interpreted as either degrees of similarity or as degrees of preference [15]. This way, fuzziness, uncertainty and also uncertainty about fuzziness are unified into a single representation. Although it is syntactically allowed, it does not make sense semantically to have an undefined value together with one or more defined values as elements of a fuzzy set of which the membership degrees are interpreted as degrees of similarity or degrees of preference. Therefore, if an undefined value is an element of such a fuzzy set, the latter is considered as (completely) undefined, i.e.,

$$\forall\, \tilde{V} \in \tilde{\wp}(dom_{ts}) : \mu_{\tilde{V}}(\bot_{ts}) \neq 0 \Rightarrow \tilde{V} = \{1/\bot_{ts}\}$$

The designated domain of a (generalised) object type remains a crisp set of object values, but its properties and components values have a generalised domain.

The domain values for components, attributes or references

$$id\ isr\ \tilde{\tilde{ts}}$$

of structured literal types and object types are constrained by the generalised constraints defined by L.A. Zadeh [4]. The considered principal types of constraints are the following:

1. *Equality constraint*, $r = e$. In this case $id\ ise\ \tilde{\tilde{ts}}$ means that id will be assigned a crisp value of $dom_{\tilde{\tilde{ts}}}$, i.e. a value of the set

   $$\{\{1/\{1/x\}\} \mid x \in dom_{ts}\}$$

   This case is semantically equivalent with the definition $id : ts$ of its crisp counterpart.
2. *Possiblilistic constraint*, $r = blank$. In this case $id\ is\ \tilde{\tilde{ts}}$ means that id will be assigned a value of $dom_{\tilde{\tilde{ts}}}$ and the membership degrees of the fuzzy sets $\tilde{V}_j$ are all interpreted as degrees of similarity, i.e. id is considered as a disjunctive (possibilistic) variable.
3. *Veristic constraint*, $r = v$. In this case, $id\ isv\ \tilde{\tilde{ts}}$ means that id will be assigned a value of $dom_{\tilde{\tilde{ts}}}$ and the membership degrees of the fuzzy

sets $\tilde{V}_j$ are all interpreted as degrees of preference, i.e. *id* is considered as a conjunctive (veristic) variable.

With the generalisation of the domains, the definitions of operators and axioms must also be extended. Within the set of extended operators $O_{\tilde{\tilde{ts}}}$ (and also within the associated set of axioms $A_{\tilde{\tilde{ts}}}$) three kinds of operators have been distinguished:

- operators of O_{ts} which remain unmodified except that their arguments are replaced by fuzzy sets representing crisp domain values, i.e. fuzzy sets of the form $\{1/\{1/x\}\}, x \in dom_{ts}$,
- operators of O_{ts} which have been generalised ("fuzzified") by applying Zadeh's extension principle twice, and
- specific operators on fuzzy sets (e.g. $\cup$, $\cap$, *co*, normalize, α-cut, $\bar{\alpha}$-cut, *core*, *supp*, ...).

The generalised type system. With these generalisations, the generalised type system GTS is defined as a triple:

$$GTS = [\tilde{\tilde{T}}, P, \tilde{\tilde{f}}^{type}_{impl}]$$

where $\tilde{\tilde{T}} = \{\tilde{\tilde{\mathbf{Void}}}\} \cup \tilde{\tilde{T}}_{reference} \cup \tilde{\tilde{T}}_{literal} \cup \tilde{\tilde{T}}_{object}$ is the generalised set of valid type specifications, P remains the infinite set of all the valid implementations and $\tilde{\tilde{f}}^{type}_{impl}$ is a mapping which maps each generalised type specification $\tilde{\tilde{ts}}$ of $\tilde{\tilde{T}}$ onto the subset $\{p_1, \ldots, p_n\}$ that contains all the implementations of $\tilde{\tilde{ts}}$:

$$\tilde{\tilde{f}}^{type}_{impl} : \tilde{\tilde{T}} \to \wp(P)$$
$$\tilde{\tilde{ts}} \mapsto \{p_1, \ldots, p_n\}$$

Instances of the generalised type system. An instance $\tilde{\tilde{t}}$ of the generalised type system GTS is called a generalised type and is defined as a couple:

$$\tilde{\tilde{t}} = [\tilde{\tilde{ts}}, \tilde{\tilde{f}}^{type}_{impl}(\tilde{\tilde{ts}})], \text{ with } \tilde{\tilde{ts}} \in \tilde{\tilde{T}}$$

Example 6. With

$$\{\tilde{\tilde{\mathbf{String}}}, \tilde{\tilde{\mathbf{Real}}}, \tilde{\tilde{\mathbf{List}}}(\tilde{\tilde{ts}}), \tilde{\tilde{\mathbf{Set}}}(\tilde{\tilde{ts}}), \tilde{\tilde{\mathbf{Enum}}}\ GTSex(male = 0, female = 1),$$
$$GTDate, \tilde{\tilde{\mathbf{Struct}}}\ GTResult(Value\ is\ \tilde{\tilde{\mathbf{Real}}}; Date\ ise\ GTDate)\} \subset \tilde{\tilde{T}}_{literal}$$

and

$$\{\tilde{\tilde{\mathbf{Set}}}(\tilde{\tilde{\mathbf{Ref}}}(\tilde{\tilde{ts}})), \tilde{\tilde{\mathbf{List}}}(\tilde{\tilde{\mathbf{Ref}}}(\tilde{\tilde{ts}}))\} \subset \tilde{\tilde{T}}_{reference}$$

a generalisation of the object types of Example 2 is given as, e.g.:

$$[GTMedication, \tilde{\tilde{f}}^{type}_{impl}(GTMedication)],\ [GTPerson, \tilde{\tilde{f}}^{type}_{impl}(GTPerson)]$$
$$\text{and } [GTPatient, \tilde{\tilde{f}}^{type}_{impl}(GTPatient)]$$

with

$$\mathbf{Cl\tilde{\tilde{a}}ss}\ GTMedication(Name\ ise\ \mathbf{Str\tilde{\tilde{i}}ng};$$
$$Composition\ ise\ \mathbf{S\tilde{\tilde{e}}t}(\mathbf{Str\tilde{\tilde{i}}ng});$$
$$Users\ is\ \mathbf{S\tilde{\tilde{e}}t}(\mathbf{R\tilde{\tilde{e}}f}(GTPatient)))$$

$$\mathbf{Cl\tilde{\tilde{a}}ss}\ GTPerson(Name\ ise\ \mathbf{Str\tilde{\tilde{i}}ng};$$
$$Given\_names\ ise\ \mathbf{L\tilde{\tilde{i}}st}(\mathbf{Str\tilde{\tilde{i}}ng});$$
$$Birth\_date\ is\ GTDate;$$
$$Sex\ ise\ GTSex;$$
$$Parents\ ise\ \mathbf{L\tilde{\tilde{i}}st}(\mathbf{R\tilde{\tilde{e}}f}(GTPerson));$$
$$Children\ ise\ \mathbf{L\tilde{\tilde{i}}st}(\mathbf{R\tilde{\tilde{e}}f}(GTPerson));$$
$$Add\_child\ ise$$
$$\mathbf{Signat}((New\_child : \mathbf{R\tilde{\tilde{e}}f}(GTPerson)) \rightarrow \mathbf{V\tilde{\tilde{o}}id}))$$

and

$$\mathbf{Cl\tilde{\tilde{a}}ss}\ GTPatient : GTPerson($$
$$Height\ is\ \mathbf{L\tilde{\tilde{i}}st}(GTResult);$$
$$Weight\ is\ \mathbf{L\tilde{\tilde{i}}st}(GTResult);$$
$$Allergies\ isv\ \mathbf{Str\tilde{\tilde{i}}ng};$$
$$Medical\_history\ ise\ \mathbf{L\tilde{\tilde{i}}st}(\mathbf{Str\tilde{\tilde{i}}ng});$$
$$Cur\_presc\ ise\ \mathbf{S\tilde{\tilde{e}}t}(\mathbf{R\tilde{\tilde{e}}f}(GTMedication));$$
$$Process\_results\ ise$$
$$\mathbf{Signat}((Results : \mathbf{L\tilde{\tilde{i}}st}(GTResult)) \rightarrow \mathbf{V\tilde{\tilde{o}}id});$$
$$Add\_child\ ise$$
$$\mathbf{Signat}((New\_child : \mathbf{R\tilde{\tilde{e}}f}(GTPatient)) \rightarrow \mathbf{V\tilde{\tilde{o}}id})) \quad \diamond$$

Instances of a generalised object type. The generalisation of a transient object results in:

$$\tilde{\tilde{o}} = [\tilde{\tilde{t}}, \tilde{\tilde{v}}, \mu_{True}, \mu_{False}]$$

whereas a generalised persistent object is defined as:

$$\tilde{\tilde{o}} = [oid, N, \tilde{\tilde{t}}, \tilde{\tilde{v}}, \mu_{True}, \mu_{False}]$$

Hereby, *oid* is still the unique object identifier, N is still a set of object names, $\tilde{\tilde{t}} = [\tilde{\tilde{ts}}, \tilde{\tilde{f}}^{type}_{impl}(\tilde{\tilde{ts}})]$ is the generalised object type, and the state $\tilde{\tilde{v}}$ is an element of $dom_{\tilde{\tilde{ts}}}$. Additionally, two elements μ_{True} and μ_{False} —which can take values between 0 and 1— are added, respectively indicating the possibility that the object is an instance of the object type and the possibility that the object is not an instance of the object type. These two elements are combined [16] into a fuzzy thruth value

$$\{\mu_{True}/True, \mu_{False}/False\}$$

of the proposition "$\tilde{\tilde{o}}$ is an instance of the generalised object type $\tilde{\tilde{t}}$". Special cases are [17]:

- $\mu_{True} = 1, \mu_{False} = 0$: the proposition is (undoubtedly) true,
- $\mu_{True} = 0, \mu_{False} = 1$: the proposition is (undoubtedly) false,
- $\mu_{True} = 1, \mu_{False} = 1$: it is not possible to decide whether the proposition is true or false,
- $\mu_{True} = 1, \mu_{False} > 0$: the proposition is true although there is some doubt about its truth,
- $\mu_{True} > 0, \mu_{False} = 1$: the proposition is false although there is some doubt about its truth.

Example 7.

$$\begin{aligned}
&[oid1, \{P721028\text{–}1101\}, GTPatient, \textbf{Struct}(\\
&\quad Name: \{1/\{1/\text{"Smith"}\}\};\\
&\quad Given_names: \{1/\{1/\textbf{List}(\{1/\{1/\text{"Ann"}\}\})\}\};\\
&\quad Birth_date: \{1/\{.8/27.10.1972, 1/28.10.1972, .8/29.10.1972\}\};\\
&\quad Sex: \{1/\{1/female\}\};\\
&\quad Parents: \perp_{\tilde{\tilde{Set}}(\tilde{\tilde{Ref}}(GTPerson))};\\
&\quad Children: \{1/\{1/\textbf{List}(\{1/\{1/oid2\}\}, \{1/\{1/oid3\}\})\}\};\\
&\quad Height: \{1/\{1/\textbf{List}(\\
&\qquad \{1/\{1/\textbf{Struct}(Value: \{\{1/\{0.8/1.70, 1/1.71, .8/1.72\}\};\\
&\qquad\qquad Date: \{1/\{1/06.04.1999\}\})\}\})\}\};
\end{aligned}$$

$$\begin{aligned}
Weigth :&\{1/\{1/\mathbf{List}(\\
&\quad \{1/\{1/\mathbf{Struct}(Value : \{1/\{1/63\}\};\\
&\qquad Date : \{1/\{1/06.04.1999\}\})\}\},\\
&\quad \{1/\{1/\mathbf{Struct}(Value : \{1/\{1/60\}\};\\
&\qquad Date : \{1/\{1/21.06.1999\}\})\}\})\}\};\\
Allergies :&\{1/\{1/\text{“penicillin”}, 0.6/\text{“cats”}\}\};\\
Medical_history :&\{1/\{1/\mathbf{List}(\{1/\{1/\text{“influenza on 02.02.1998”}\}\})\}\};\\
Cur_presc :&\{1/\{1/\mathbf{Set}(\ )\}\}), 1, 0]
\end{aligned}$$

is an example of a persistent object that is certainly an instance of the generalised type *GTPatient* ⋄

3.2 Generalised Constraint System

The generalisation of the constraint system is performed as follows: each Boolean (constraint) function has been generalised to a function $\tilde{\tilde{c}}$ with co-domain $dom_{\tilde{\tilde{Boolean}}}$. The result of this function defines the fuzzy truth value of the proposition " $\tilde{\tilde{o}}$ satisfies the constraint $\tilde{\tilde{c}}$ " as:

$$\{\mu_{\tilde{\tilde{c}}(\tilde{\tilde{o}})}(\{1/\mathbf{True}\})/True, \mu_{\tilde{\tilde{c}}(\tilde{\tilde{o}})}(\{1/\mathbf{False}\})/False\}$$

which indicates that $\tilde{\tilde{o}}$ satisfies the constraint $\tilde{\tilde{c}}$ with a possibility

$$\mu_{\tilde{\tilde{c}}(\tilde{\tilde{o}})}(\{1/\mathbf{True}\})$$

and does not satisfy the constraint $\tilde{\tilde{c}}$ with a possibility

$$\mu_{\tilde{\tilde{c}}(\tilde{\tilde{o}})}(\{1/\mathbf{False}\})$$

All the constraint specifications of the crisp constraint system CS are generalised in this way, resulting in the generalised sets of constraint specifications $\tilde{\tilde{C}}^{single}_{instance}$, $\tilde{\tilde{C}}^{single}_{extent}$, $\tilde{\tilde{C}}^{multi}_{instance}$ and $\tilde{\tilde{C}}^{multi}_{extent}$.

The generalised constraint system GCS is then defined as:

$$GCS = [\tilde{\tilde{C}}, P, \tilde{\tilde{f}}^{constr}_{impl}]$$

where $\tilde{\tilde{C}} = \tilde{\tilde{C}}^{single}_{instance} \cup \tilde{\tilde{C}}^{single}_{extent} \cup \tilde{\tilde{C}}^{multi}_{instance} \cup \tilde{\tilde{C}}^{multi}_{extent}$ is the generalised set of valid constraint specifications, P is the infinite set of all programs and $\tilde{\tilde{f}}^{constr}_{impl}$ denotes the generalised (constraint) implementation function:

$$\begin{aligned}
\tilde{\tilde{f}}^{constr}_{impl} : \tilde{\tilde{C}} &\to \wp(P)\\
\tilde{\tilde{cs}} &\mapsto \{p_1, \ldots, p_n\}
\end{aligned}$$

$\tilde{\tilde{f}}^{constr}_{impl}$ is a mapping which maps each generalised constraint specification $\tilde{\tilde{cs}}$ of the domain $\tilde{\tilde{C}}$ onto its set $\{p_1, \ldots, p_n\}$ of implementations.

Instances of the generalised constraint system. An instance $\tilde{\tilde{c}}$ of the generalised constraint system GCS is defined as a couple:

$$\tilde{\tilde{c}} = [\tilde{\tilde{cs}}, \tilde{\tilde{f}}^{constr}_{impl}(\tilde{\tilde{cs}})], \text{ with } \tilde{\tilde{cs}} \in \tilde{\tilde{C}}$$

3.3 Generalised Object Schemes and Generalised Database Schemes

The definitions of a generalised object scheme and of a generalised database scheme are the generalisations of the definitions given in Subsection 2.3.

A generalised object scheme and its instances. A generalised object scheme $\tilde{\tilde{os}}$ is defined as a triple

$$\tilde{\tilde{os}} = [\tilde{\tilde{t}}, \tilde{\tilde{M}}, \tilde{\tilde{C}}_{\tilde{\tilde{t}}}]$$

where

- the first component $\tilde{\tilde{t}}$ is a generalised object type (i.e.

  $$\tilde{\tilde{t}} = [\tilde{\tilde{ts}}, \tilde{\tilde{f}}^{type}_{impl}(\tilde{\tilde{ts}})]$$

  is an instance of the generalised type system GTS and $\tilde{\tilde{ts}} \in \tilde{\tilde{T}}_{object}$),
- $\tilde{\tilde{M}}$ is an informal component which is described in natural language and represents the meaning of the generalised object scheme and
- $\tilde{\tilde{C}}_{\tilde{\tilde{t}}}$ is a fuzzy set of instances of the generalised constraint system GCS, whose membership degrees are interpreted as degrees of preference [15] expressing the relevance of the constraints. Each instance of $\tilde{\tilde{C}}_{\tilde{\tilde{t}}}$ is single-type dependent and applies to a single instance of the generalised object type $\tilde{\tilde{t}}$, i.e.

  $$\tilde{\tilde{C}}_{\tilde{\tilde{t}}} \in \tilde{\wp}(\tilde{\tilde{C}}^{single}_{instance})$$

  The membership degree of an element of $\tilde{\tilde{C}}_{\tilde{\tilde{t}}}$ indicates to which degree the constraint applies to the object type $\tilde{\tilde{t}}$.

The set of instances of a generalised object scheme $\tilde{\tilde{os}} = [\tilde{\tilde{t}}, \tilde{\tilde{M}}, \tilde{\tilde{C}}_{\tilde{\tilde{t}}}]$ is by definition the set of all instances of the object type $\tilde{\tilde{t}}$ that satisfy all constraints of $\tilde{\tilde{C}}_{\tilde{\tilde{t}}}$ and all constraints of the sets $\tilde{\tilde{C}}_{\hat{\tilde{\tilde{t}}}}$ of the object schemes $[\hat{\tilde{\tilde{t}}}, \widehat{\tilde{\tilde{M}}}, \tilde{\tilde{C}}_{\hat{\tilde{\tilde{t}}}}]$ which have been defined for the supertypes $\hat{\tilde{\tilde{t}}}$ of $\tilde{\tilde{t}}$.

The element μ_{True} (resp. μ_{False}) of an instance $\tilde{o}$ is obtained by a fuzzy aggregation of the membership degrees $\mu_{\tilde{\tilde{c}}_i(\tilde{o})}(\{1/\mathbf{True}\})$ (resp. $\mu_{\tilde{\tilde{c}}_i(\tilde{o})}(\{1/\mathbf{False}\})$) of the fuzzy truth values of all constraints $\tilde{\tilde{c}}_i$ it must satisfy. The choice of an aggregation function depends on how stringent the user wants the constraints to be applied. As an example a weighted average could be chosen using the relevance of the constraints as weights.

Example 8. With respect to the object types

$$[GTPerson, \tilde{f}^{type}_{impl}(GTPerson)] \text{ and } [GTPatient, \tilde{f}^{type}_{impl}(GTPatient)]$$

of Example 2, the object schemes $GOSPerson$ and $GOSPatient$ can be constructed as:

$$GOSPerson = [GTPerson, \tilde{\tilde{M}}, \tilde{\tilde{C}}_{GTPerson}]$$

where:

- $GTPerson$ is the (generalised) type of the generalised object scheme;
- the meaning $\tilde{\tilde{M}}$ is given as:

  "This generalised object scheme is used as the specification of an abstraction for persons, such as they appear in the real world.";
- and the set $\tilde{\tilde{C}}_{GTPerson}$ is constructed as:

  $$\begin{aligned}&\tilde{\tilde{C}}_{GTPerson} = \{\\&\tilde{\tilde{\rho}}^{not\_null}_{\{GTPerson.Name\}}[\,],\\&\tilde{\tilde{\rho}}^{value}_{\{GTPerson.Birth\_date\}}[GTPerson.Birth\_date\ is\_a\ valid\_date\_of\_birth]\}\end{aligned}$$

  where "the_last_130_years" is the fuzzy set which is represented in Figure 6.

and

$$GOSPatient = [GTPatient, \tilde{\tilde{M}}, \tilde{\tilde{C}}_{GTPatient}]$$

where:

- $GTPatient$ is the type of the object scheme;
- the meaning $\tilde{\tilde{M}}$ is given as:

  "This generalised object scheme is used as the specification of a simplified abstraction of CPR's as they exists in the cabinet of a physician."

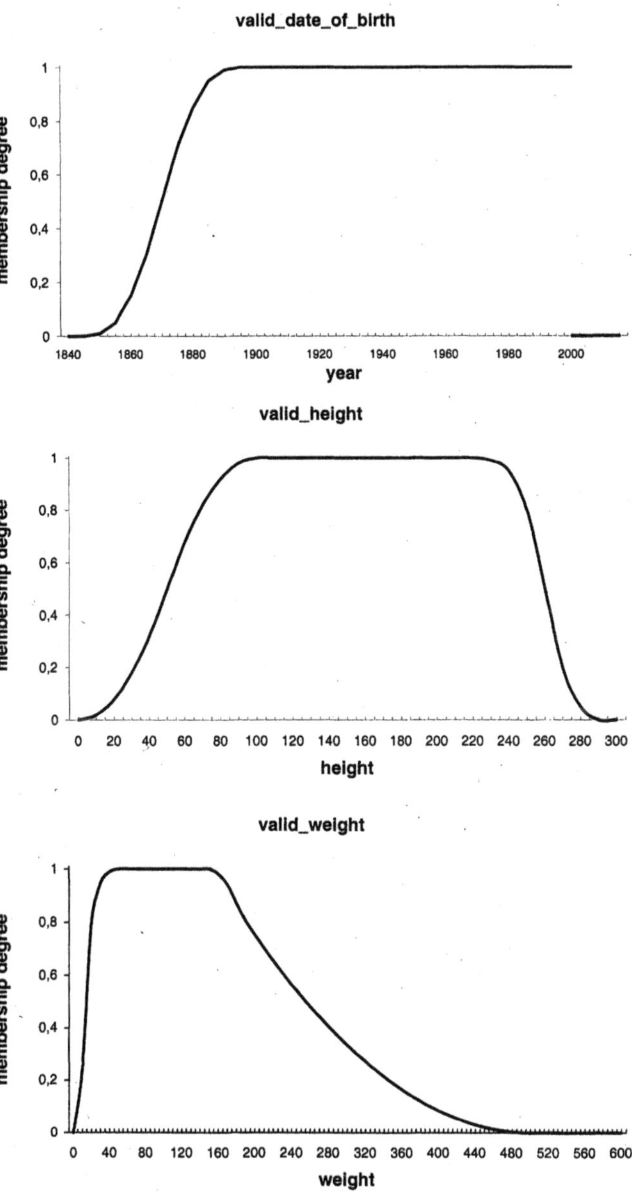

Fig. 6. The representation of the used fuzzy sets

- and the set $\tilde{\tilde{C}}_{GTPatient}$ is constructed as:

$$\tilde{\tilde{C}}_{GTPatient} = \{$$
$$\tilde{\tilde{\rho}}^{not\_null}_{\{GTPatient.PatientID\}}[\,],$$
$$\tilde{\tilde{\rho}}^{value}_{\{GTPatient.Height.Value\}}[$$
$$GTPatient.Height.Value \; is\_a \; valid\_height],$$
$$\tilde{\tilde{\rho}}^{value}_{\{GTPatient.Weight.Value\}}[$$
$$GTPatient.Weight.Value \; is\_a \; valid\_weight],$$
$$\tilde{\tilde{\rho}}^{trans}_{\{GTPatient.Allergies\}}[$$
$$(GTPatient.Allergies)^{new} \supseteq GTPatient.Allergies]\} \quad \diamond$$

where "valid_heights" and "valid_weights" are the fuzzy sets which are represented in Figure 6.

A generalised database scheme and its instances. Similarly, a generalised database scheme $\tilde{\tilde{ds}}$ describes the information which is stored in the generalised database and is defined as a triple:

$$\tilde{\tilde{ds}} = [\tilde{\tilde{D}}, \tilde{\tilde{M}}, \tilde{\tilde{C}}_{\tilde{\tilde{D}}}]$$

where

- $\tilde{\tilde{D}} = \{\tilde{\tilde{os}}_i = [\tilde{\tilde{t}}_i, \tilde{\tilde{M}}_i, \tilde{\tilde{C}}_{\tilde{\tilde{t}}_i}] | i = 1, 2, \ldots, n, n \in \mathbb{N}_0\}$ is a finite set of generalised object schemes. Every generalised object scheme in $\tilde{\tilde{D}}$ is defined for a different generalised object type. If a generalised object scheme $\tilde{\tilde{os}}_i \in \tilde{\tilde{D}}$ is defined for a generalised object type $\tilde{\tilde{t}}$ and $\tilde{\tilde{t}}'$ is a supertype of $\tilde{\tilde{t}}$ or if there exists a reference between $\tilde{\tilde{t}}'$ and $\tilde{\tilde{t}}$, then there exists a generalised object scheme $\tilde{\tilde{os}}'_i \in \tilde{\tilde{D}}$ which is defined for $\tilde{\tilde{t}}'$;
- $\tilde{\tilde{M}}$ is an informal component which is described in natural language and represents the meaning of $\tilde{\tilde{ds}}$ and
- $\tilde{\tilde{C}}_{\tilde{\tilde{D}}}$ is a fuzzy set of instances of the generalised constraint system GCS, whose membership degrees are interpreted as degrees of preference [15] expressing the relevance of the constraints. Each instance is an element of $\tilde{\tilde{C}}^{single}_{extent} \cup \tilde{\tilde{C}}^{multi}_{instance} \cup \tilde{\tilde{C}}^{multi}_{extent}$ and is defined for either the instances or the extent of the generalised object type $\tilde{\tilde{t}}_i$ of a generalised object scheme $\tilde{\tilde{os}}_i \in \tilde{\tilde{D}}$. This fuzzy set imposes extra conditions on the instances of the set of generalised object schemes. For every generalised object scheme $\tilde{\tilde{os}}_i \in \tilde{\tilde{D}}$, there exist uniqueness constraints in $\tilde{\tilde{C}}_{\tilde{\tilde{D}}}$ which guarantee the uniqueness of the object identifiers and object names of the instances of

$\tilde{\tilde{os}}_i$. Again, the membership degrees are an indication for the degree to which the constraints apply.

The set of instances of a generalised database scheme $\tilde{\tilde{ds}} = [\tilde{\tilde{D}}, \tilde{\tilde{M}}, \tilde{\tilde{C}}_{\tilde{\tilde{D}}}]$ is by definition the set of all persistent instances of the generalised object schemes $\tilde{\tilde{os}}_i \in \tilde{\tilde{D}}$ that satisfy all constraints of $\tilde{\tilde{C}}_{\tilde{\tilde{D}}}$.

As in the case of the instances of a generalised object scheme, the element μ_{True} (resp. μ_{False}) of an instance $\tilde{\tilde{o}}$ is obtained by a fuzzy aggregation of the membership degrees $\mu_{\tilde{\tilde{c}}_i(\tilde{\tilde{o}})}(\{1/\mathbf{True}\})$ (resp. $\mu_{\tilde{\tilde{c}}_i(\tilde{\tilde{o}})}(\{1/\mathbf{False}\})$) of the fuzzy truth values of all constraints $\tilde{\tilde{c}}_i \in \tilde{\tilde{C}}_{\tilde{\tilde{D}}}$.

The extent of a generalised database scheme $\tilde{\tilde{ds}}$ is called a (fuzzy) database and consists of the set of all instances of that generalised database scheme. These can be grouped into sets of instances of the same generalised object scheme.

4 Conclusion

A formal framework for the definition of a fuzzy and/or uncertain object-oriented database model has been presented. This framework is based on a type system and a related constraint system, which is meant to guarantee database integrity. With the ODMG specifications as a starting-point, we developed a formal type system. Next, we added a constraint system consisting of a variety of constraint types, elaborated in detail in this paper. Based on generalised constraints, as defined by L.A. Zadeh, the formalism is generalised in such a way that it supports both fuzziness and uncertainty. The resulting framework still completely supports the ODMG specifications, whereby the generalisation reflects the directions of the UFO-model.

References

1. De Caluwe, R. (ed.): Fuzzy and Uncertain Object-oriented Databases: Concepts and Models. Advances in Fuzzy Systems - Applications and Theory **13** (1997) World Scientific Publishing Co., Singapore
2. Cattell, R. G. G., Barry, D. K. (eds.): The Object Database Standard: ODMG 2.0. (1997) Morgan Kaufmann Publishers Inc., San Francisco CA USA
3. Alagić, S.: The ODMG Object Model: Does it Make Sense? ACM SIGPLAN Notices **32 10** (1997) 253–270
4. Zadeh, L. A.: Toward a theory of fuzzy information granulation and its centrality in human reasoning and fuzzy logic. Fuzzy Sets and Systems **90 2** (1997) 111–127
5. Van Gyseghem, N., De Caluwe, R.: The UFO Database Model: Dealing with Imperfect Information. Fuzzy and Uncertain Object-Oriented Databases: Concepts and Models, De Caluwe, R. (ed.) Advances in Fuzzy Systems - Applications and Theory **13** (1997) World Scientific Publishing Co., Singapore 123–177

6. Abiteboul, S., Kanellakis, P., Waller, E.: Method schema's. Proceedings of the 9th ACM PODS Conference (1990) ACM Press, New York NY USA 16–27
7. Cluet, S., Delobel, C., Lécluse, C., Richard, P.: Reloop, an Algebra Based Query Language for an Object-Oriented Database System. Deductive and Object-Oriented Databases, Kim, W., Nicolas, J.-M., Nishio, S. (eds.) (1990) Elsevier Science Publishers B.V., Noord Holland Amsterdam 313–331
8. Su, S. Y. W., Guo, M., Lam, H.: Association Algebra: A Mathematical Foundation for Object-Oriented Databases. IEEE Transactions on Knowledge and Data Engineering **5 5** (1993) 775–798
9. Kifer, M., Lausen, G., Wu, J.: Logical Foundations of Object-Oriented and Frame Based Languages, Journal of the ACM **42 4** (1995) 741–843
10. Fegaras, L., Maier, D., Towards an Effective Calculus for Object Query Languages. ACM SIGMOD Record **24 2** (1995) 47–58
11. Riedel, H., Scholl, M. H.: A Formalization of ODMG Queries. Proceedings of the 7th Working Conference on Database Semantics (DS-7), Spaccapietra, S., Maryanski, F. (eds.) (1997) IFIP 2.6 Chapman & Hall, Leysin Switserland 63–90
12. Lausen, G., Vossen, G.: Models and Languages of Object-Oriented Databases. International Computer Science Series (1998) Addison-Wesley, Harlow England
13. Earley, J.: Toward an understanding of data structures. Communications of the ACM **14 10** (1971) 617–627
14. Paredaens, J., De Bra, P., Gyssens, M., Van Gucht, D.: The Structure of the Relational Database Model. EATCS: Monographs on Theoretical Computer Science **17**, Brauer, W., Rozenberg, G., Salomaa, A. (eds.) (1989) Springer-Verlag, Heidelberg Germany
15. Dubois, D., Prade, H.: The three semantics of fuzzy sets. Fuzzy Sets and Systems **90 2** (1997) 141–150
16. Van Schooten A., De Caluwe, R., Kerre, E.: An expert system Development Tool Using Fuzzy Logic. Intelligent Autonomous Systems, Hertzberger, L. (ed.) (1986) Elsevier, Amsterdam the Netherlands 668–677
17. De Caluwe, R., Vandenberghe, R., Van Gyseghem, N., Van Schooten, A.: Integrating Fuzziness in Database Models. Fuzziness in Database Management Systems, Bosc, P., Kacprzyk, J. (eds.) (1995) Physica-Verlag, Heidelberg Germany, 71–113

An Implementation for Fuzzy Deductive Relational Databases

Ignacio Blanco, Juan C. Cubero, Olga Pons, and Amparo Vila

Department of Computer Science and Artificial Intelligence, University of Granada, Granada, Spain

Abstract. This chapter shows how to integrate the representation of deductive rules and fuzzy information stored in a relational DBMS to build a module that can obtain new data from data stored in tables. The deductions can be applied to classical (or precise) data, imprecise data or both of them, so it is necessary to provide a mechanism to find the tuples in the database satisfying a rule, i.e. a mechanism to calculate the precision degree of the answer by means of the combination of the precision degrees of every value into an unified measure.
Keywords. relational databases extension, fuzzy deduction, inference.

1 Introduction

Since the relational model was introduced by Codd in [1], it has been used in a generalized way for the representation and handling of imprecise and "fuzzy" information by many authors [2–6].

In such a model, we store information in tables. For instance, we can store information about the age of people in a table, and we can recover it if we want to find out the age of a person or an interval of age. But this type of information is not usual when a human being reasons.

We usually ask for "young" or "old" people without considering that such concepts vary from one person to another, and that we can not assure who belongs to the set of people satisfying this property and who does not. When we want to find people that are twenty-four, the query result is a set of people made up of only those people who are twenty-four.

In these cases, there are two things we can do: provide a method to introduce and query the flexible values, or provide a flexible query language to make flexible queries on precise values. Medina, Pons and Vila integrated the two ideas when they presented GEFRED [7].

This was the "flexible", "imprecise" or as it is called "Fuzzy", part of the problem of approaching human reasoning. But there is another problem.

When a human being thinks, he obtains data from other data, i.e. he is able to deduce information that depends on itself. For instance, data about a person's father, mother and ancestors can be stored in separate tables, but ancestors can be obtained from the father and mother tables queried in a fixed order, i.e. a person's ancestor is a person's father and mother, the father and mother's father and mother and so on.

Another aspect to be considered is the one derived when storing explicitly information that is implicitly stored in database. If source data are modified, implicit information in database is automatically modified. But if implicit information is stored explicitly in database, instead of creating an algorithm to calculate it, the modification of source data has to be explicitly stored.

When we obtain data from other data, we follow a set of steps or "rules" that we could represent in a logical structure and we could design an algorithm that uses that structure to access the tables to complete the dependent table.

To deal with this, we would have a representation of the flexible values and a language for querying them, a flexible query language for querying concrete values and a structure to represent the logical rules, and an algorithm to deduce data from other ones.

Finally, we could integrate all these in a single system so that we make deductions on flexible or concrete data.

2 Theoretical background

This work is not intended to be a theoretical one but rather an implementation of Medina, Pons, Vila and Cubero's FREDDI [8].

The authors gave the name FSQL to Fuzzy SQL, i.e. the flexible query language on flexible and precise values, and gave the name DFSQL to Deductive FSQL, i.e. the query language that allows deductions to be made on flexible and precise values.

As they set out, the structure for a server with these capabilities is shown in underneath figure 1.

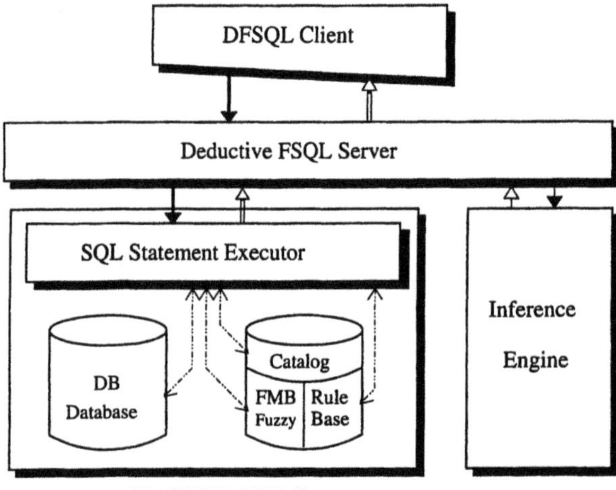

Fig. 1. Original DFSQL system structure in FREDDI

The functions of each component appearing in figure 1 are the following:

DFSQL Client: it accepts DFSQL sentences from the user, sends them to the DFSQL Server which returns and formats the query result so that it can be presented to the user.

DFSQL Server: it accepts DFSQL sentences sent by DFSQL Clients, separates parts involving fuzzy treatment and intensional tables and treats them separately. We will distinguish between FSQL, DSQL (Deductive SQL) and DFSQL sentences. All of them are translated to a set of classical SQL sentences and procedures and function calls.

SQL Statement Executor: it accepts classical SQL sentences in which the DFSQL Server has translated user DFSQL sentences and executes them on the database, using data stored in the Classical Catalog and Extended Catalog integrated by the Fuzzy Meta-base and Rule Base.

Database: it contains the user data.

Catalog: it contains the information about the user data stored in the database, i.e. the so-called Meta-data.

Extended Catalog: it contains the information about the fuzzy data stored in the database (FMB or Fuzzy Meta-Base) and about the logical rules used to infer about the data stored in the database (Rules Base).

Inference Engine: it uses logical rules stored in the Rule Base to infer new data using data stored in the Database and in the Extended Catalog.

3 Implementation aspects and original model modifications

We have to consider that when we extend a Relational DataBase Management System (RDBMS) to give it fuzzy and deductive capabilities, we are restricted by the relational model and all structures have to be represented using such a model.

3.1 Concrete implementation

FREDDI was created using Oracle©RDBMS as a reference, so we will use it too.

At present, fuzzy and deductive capabilities are implemented in two separate servers with no communication between the two. But we will later explain how the integration module of the two servers which we are developing operates.

Up until now, Oracle©has incorporated new features that introduce some variations on the original architecture of FREDDI. Originally, a Prolog Engine was the inference engine and it was a built-out module with respect to the RDBMS, implemented in a programming language that had to communicate with Oracle©or such communication had to be programmed separately.

Nowadays, Oracle©provides programming languages that have built-in communication capabilities.

There are two possibilities which depend on the Oracle©provided language chosen:

- The first option is Pro*C©. This is a host language, i.e. a language (C) that incorporates extended sentences for using or communicating with Oracle©. It presents some handicaps and the most important one is that it requires a network to send messages either that it can communicate with Oracle©.
- The second is PL/SQL©. This is a programming database language built in the RDBMS. The modules created with it are stored in the database, and they are executed by the RDBMS directly, so they do not need a network. But this language is less general and more restrictive than the first.

3.2 Model modifications

Depending on the language chosen, we will have some variations of the basic architecture.

Deductive built-out engine The architecture suggested when a language like Pro*C©is used, i.e. a host language that is external to the RDBMS, is shown in the underneath figure 2

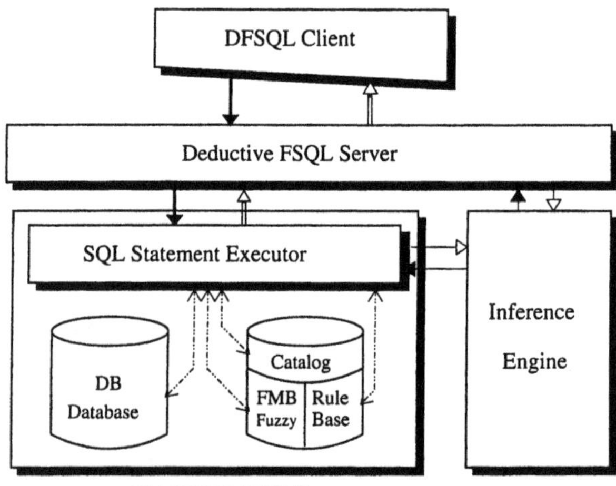

Fig. 2. Built-out deductive engine architecture

Deductive built-in engine The architecture suggested when a language like PL/SQL©is used, i.e. a built-in RDBMS language, is shown in the figure 3

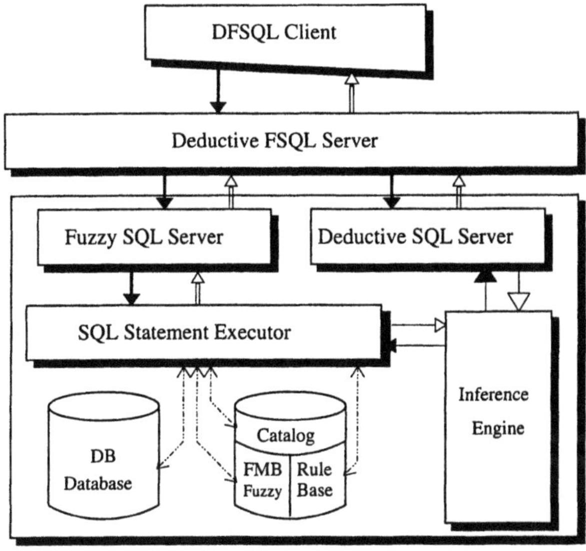

Fig. 3. Built-in deductive engine architecture

4 Representing fuzzy and logical information in the relational model

As has been previously stated, the Extended Catalog stores information about the fuzzy data and logical rules stored in the database, but it is divided into two sub-catalogs:

Fuzzy Meta-Base or FMB: stores information about fuzzy data.
Rule Base: stores logical rules to make deductions.

4.1 Fuzzy data catalog: FIRST

When we answer a question, in many cases our answer is not exact, i.e. human beings often use **imprecision**. We are imprecise when we give an approximate response to a question or when we know nothing about the answer. It implies that there are several types of imprecision:

- when the information we have is not complete,

- when we know nothing about the information certainty (**uncertainty**),
- when we do not know the information or
- when the information requested is not applicable to the entity.

The types of fuzzy attributes we can represent in our system are:

- *Fuzzy type 1 attributes* are attributes that can take crisp values but that can be queried by specifying fuzzy conditions on their values expressed by labels defined on their domain.
- *Fuzzy type 2 attributes* are attributes that can store fuzzy numbers, a concept that was introduced by Zadeh in [9]. These fuzzy numbers are represented by possibility distributions defined on continuous or discrete domains in which an order relation exists. Possible values for this type of attributes are:
  - *Trapezoid-shaped possibility distribution* that is a representation of the membership function associated to data and is given by four parameters $[\alpha, \beta, \gamma, \delta]$ (note we are using normalized fuzzy values).
  - *Linguistic label* that references a linguistic concept subjectively defined and with an associated possibility distribution, for instance, label "tall" for the height of a person.
  - *Approximate value* represents the fuzzy concept "approximately n" where n is a value of the ordered underlying domain. The possibility distribution for this fuzzy value is a triangle-shaped one determined by a parameter called *margin* and represented by $[n-margin, n, n, n+margin]$.
  - *Possibility intervals* are special trapezoid-shaped possibility distributions where there is no rising and fading lines on the trapezoid sides. This value was proposed by Grant [10] and is represented by $[\alpha, \alpha, \beta, \beta]$.
- *Fuzzy type 3 attributes* are attributes that are defined on discrete non-ordered domains where a similarity relation between values exists. Possible values of this type are:
  - *Scalar value* is a possibility distribution where only the possibility value 1 exists. The possibility distribution is $(1, d)$, i.e. only the scalar value d can appear and its possibility value is 1.
  - *Possibility distribution defined on a scalar domain:* a datum is assigned a possibility distribution where all underlying domain values are scalar and not ordered $\{(p_1, d_1), \ldots, (p_n, d_n)\}$.

For all these fuzzy types, there are special values introduced by Codd [11–14] that must be considered:

- *UNKNOWN* represents ignorance of an attribute value. It implies that an attribute can take any value in the domain with a possibility of 1 so the distribution representing this value is $\{1/u, \forall u \in U\}$ where U is the underlying domain.

- *UNDEFINED* is when there is no possible value for this attribute in this tuple so the distribution representing this value is $\{0/u, \forall u \in U\}$ where U is the underlying domain.
- *NULL* is when no information about the attribute value is supplied because it is either UNKNOWN or UNDEFINED (we don't know which). The distribution possibility is $\{1/\text{UNKNOWN}, 1/\text{UNDEFINED}\}$.

To summarize, what we have to store for fuzzy type 2 values is shown in table 1 and for fuzzy type 3 values in table 2.

Value Types	DB Attributes for Fuzzy Type 2 Values				
	FT	F1	F2	F3	F4
UNKNOWN	0	NULL	NULL	NULL	NULL
UNDEFINED	1	NULL	NULL	NULL	NULL
NULL	2	NULL	NULL	NULL	NULL
CRISP	3	d	NULL	NULL	NULL
LABEL	4	FUZZY_ID	NULL	NULL	NULL
INTERVAL$[n, m]$	5	n	NULL	NULL	m
APPROX(d)	6	d	$d - margin$	$d + margin$	$margin$
TRAPEZOID$[\alpha, \beta, \gamma, \delta]$	7	α	β	γ	δ

FT stores value type and *F*1, *F*2, *F*3 and *F*4 are used to store each value parameters. Values NULL appearing in this table means an *inapplicable value* in the host RDBMS.

Table 1. Internal representation in the host RDBMS for fuzzy type 2 attributes

Value Types	DB Attributes for Fuzzy Type 3 Values					
	FT	FP1	F1	...	FPn	Fn
UNKNOWN	0	NULL	NULL	...	NULL	NULL
UNDEFINED	1	NULL	NULL	...	NULL	NULL
NULL	2	NULL	NULL	...	NULL	NULL
SIMPLE	3	p	d	...	NULL	NULL
POSS	4	p_1	d_1	...	p_n	d_n

FT stores the value type and pairs (p_n, d_n) represent that the domain value d_n has the possibility value p_n.

Table 2. Internal representation in host RDBMS for fuzzy type 3 attributes

The table structure for the FMB is shown in figure 4 and it is based on the Oracle©RDBMS structure.

We will describe briefly the information contained in each table:

- *FUZZY_COL_LIST*: contains the list of attributes in the database that can be treated in a fuzzy way. Each attribute is described by the table

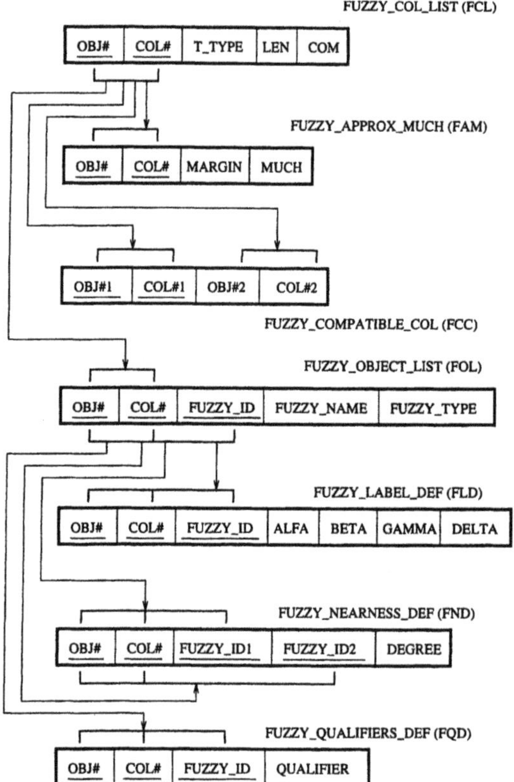

Fig. 4. FMB tables structure

to which it belongs(OBJ#) and the identifier referencing it in the table (COL#), its fuzzy type (F_TYPE), number of values in the possibility distribution if the attribute is fuzzy type 3 (LEN) and a comment about the attribute (COM).

- *FUZZY_OBJECT_LIST*: stores fuzzy objects that represent values for the database attributes. Each object is described by the table and column containing it, an identifier, a name and a type.
- *FUZZY_LABEL_DEF*: contains information about the trapezoid-shaped possibility distributions that are associated to the linguistic labels. Each one is described by the table and column containing the associated label, the associated label identifier and α, β, γ and δ parameters.
- *FUZZY_APPROX_FCOMP*: contains *margin* and *fcomp* parameters used if working on fuzzy type 1 or 2 objects. Two parameters (MARGIN, MUCH) are associated to a column of a table (OBJ#,COL#).
- *FUZZY_NEARNESS_DEF*: contains similarity values between value pairs in the same domain. These similarity relations are allowed between fuzzy type 3 values. These degrees (DEGREE) are associated to value pairs

(FUZZY_ID1,FUZZY_ID2) that can appear in a fuzzy column of a table (OBJ#,COL#).

- *FUZZY_COMPATIBLE_COL*: contains information about the compatibility of fuzzy type 3 attributes so if an attribute is compatible to another one it is not necessary to define all the labels and similarity relations again. Each row represents a relation of compatibility between an object (OBJ#1,COL#1) and another one (OBJ#2,COL#2).
- *FUZZY_QUALIFIERS_DEF*: contains minimum satisfying thresholds for each qualifier (QUALIFIER) of a linguistic label (FUZZY_ID) that can be contained in a column of a table (OBJ#,COL#).

Fuzzy Meta-Base description is deeply presented in [15].

These tables are only used by the DFSQL Server owned by the DB administrator, and all clients access them through user views not allowing these tables to be modified.

The syntax of the sentences for fuzzy catalog management is shown in appendix A.

4.2 Deductive data catalog: DEDUCTOR

Let us imagine a table like the one shown in table 3, and let us suppose that it is necessary to find out information about a person's ancestors. There are two possibilities. The first one is to represent this information in an extensional way in the database as shown in table 4.

Father/Mother	**Son/Daughter**
John	Mike
Mary	Mike
Karl	Louise
Maggie	Louise
Mike	Jake
Louise	Jake

Table 3. Table PARENTS

The addition of new entries to the table PARENTS implies adding several new entries to the table ANCESTORS. In many cases, the number of entries added to the table ANCESTORS is great.

However, there is another way to obtain information about the ancestors and that is by trying to represent the way in which the table ANCESTOR is completed.

It is known that a parent is an ancestor, and a parent of parents is an ancestor, and so on. If we represent it using logical rules:

Father/Mother	Son/Daughter
John	Mike
John	Jake
Mary	Mike
Mary	Jake
Karl	Louise
Karl	Jake
Maggie	Louise
Maggie	Jake
Mike	Jake
Louise	Jake

Table 4. Table ANCESTORS

$$ANCESTOR(X,Y) \Longleftarrow PARENT(X,Y) \tag{1}$$
$$ANCESTOR(X,Y) \Longleftarrow PARENT(X,Z)\ AND\ ANCESTOR(Z,Y)$$

If a mechanism to obtain data from these rules can be designed, the table ANCESTOR can be updated automatically when new entries are added to the table PARENT.

Two types of predicates can be seen in the rules below. The predicate PARENT is associated to a table which contains data, but the predicate ANCESTOR is not. So these types must be considered:

- *Extensional* predicates are those associated to extensional tables, i.e. those whose table content does not depend on the content of other tables. Each variable in an extensional predicate is referred to a column in the related extensional table. This correspondence is stored in the catalog table CONVERSION.
- *Intensional* predicates are those associated to intensional tables, i.e. those whose table content depends on the content of other tables. These intensional tables and predicates are registered in a catalog table called INTENSIONAL_CATALOG.

The basic case in the deduction are level one depth rules that are rules which only contain extensional or pre-calculated intensional predicates. But it is not usually the case. Instead of this, we can find level greater-than-one depth rules that contain intensional not previously calculated predicates that involve new deductions. And the last case is level greater-than-one depth rules that contain the predicate being defined by the rule, i.e. *recursive* case.

A deduction rule has the format:

$$P \leftarrow K_1 \vee K_2 \vee \ldots \vee K_m \wedge K_{m+1} \wedge K_{m+2} \wedge \ldots \wedge K_n \tag{2}$$

If all disjunctive clauses are put on one side, and all the conjunctive ones on the other, the rules are:

$$P \leftarrow (K_1 \vee K_2 \vee \ldots \vee K_m) \wedge (K_{m+1} \wedge K_{m+2} \wedge \ldots \wedge K_n) \tag{3}$$

If distribution is applied, the rules are:

$$P \leftarrow (K_1 \wedge (K_{m+1} \wedge K_{m+2} \wedge \ldots \wedge K_n)) \vee (K_2 \wedge (K_{m+1} \wedge K_{m+2} \wedge \wedge \ldots \wedge K_n)) \vee \ldots \vee (K_m \wedge (K_{m+1} \wedge K_{m+2} \wedge \ldots \wedge K_n)) \tag{4}$$

that is, rule is finally defined by:

$$P \leftarrow P_1 \vee P_2 \vee \ldots \vee P_m \tag{5}$$

where

$$P_i \leftarrow K_i \wedge (K_{m+1} \wedge K_{m+2} \wedge \ldots \wedge K_n) \tag{6}$$

So a rule can be decomposed in disjunction for rules which only contain a conjunction operator. A possible tables structure for representing this information in an RDBMS is shown in figure 5.

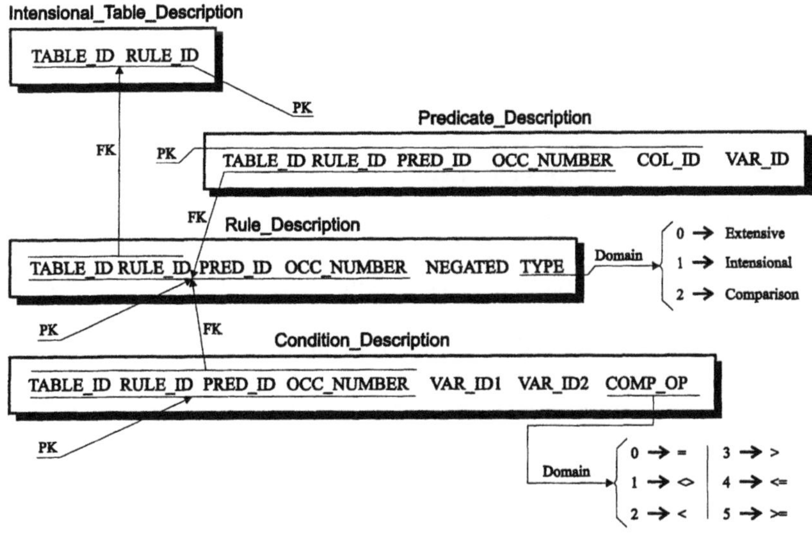

Fig. 5. Rule base tables structure

Let us now look at a brief description of each table:

- *INTENSIONAL_TABLE_DESCRIPTION*: stores intensional predicates P (TABLE_ID) defined like the disjunction of rules P_i (RULE_ID).

- *RULE_DESCRIPTION*: describes each rule P_i as a sequence of extensional or intensional predicates or comparisons joined by conjunction. The predicates can appear negated. The same predicate can appear more than once in a rule but each occurrence is identified by its position. TABLE_ID and RULE_ID identify the rule, PRED_ID the predicate, OCC_NUMBER the position within the rule, NEGATED indicates if the predicate is negated and TYPE id whether the predicate is extensional, intensional or comparison.
- *PREDICATE_DESCRIPTION*: describes the order of variables in each predicate K_i. The same variable can appear more than once in a predicate but several occurrences of the same one are distinguished by their positions in the predicate. TABLE_ID and RULE_ID identify the rule, PRED_ID the predicate, OCC_NUMBER indicates the predicate position within the rule, VAR_ID identifies the variable and COL_ID indicates the position within the predicate.
- *CONDITION_DESCRIPTION*: describes the conditions, the special type of predicates, that have only two variables and its type is one of the following: $=$, $\neq$, $\leq$, $<$, $\geq$ and $>$. TABLE_ID and RULE_ID identify the rule, PRED_ID the predicate, OCC_NUMBER indicates the predicate position within the rule, VAR_ID1 identifies the first variable in the condition and VAR_ID2 the second one, and COMP_OP indicates comparison type.

The syntax of the sentences for deductive catalog management is shown in appendix B.

5 Handling logical information

5.1 Basis: Prolog-like mechanism

There is a classical logical language that represents this type of logical rules and uses them to make deductions. We have used the intuitive idea on which the Prolog mechanism deduction is based. This idea is one of fixing the value of variables.

The result of a rule expansion is a set of variable value tuples that satisfy the rule. Some of these variables correspond to defined predicate variables and the others are internal to the rule variables used to connect predicates in the rule body. Let us illustrate this with an example based on the rules expressed in the equation 2. The variable value sets for the second rule would have the structure:

X	Y	Z

The basic structure of the value set will be copied to represent all the possible results of the rule expansion. To store this information, we will use a table called *PRED<pred>_RULE<rule>* where <pred> represents the predicate identifier defined by the rule and <rule> the rule identifier. The structure for these tables is:

IDSET	VAR1	...	VARn	NEG	INST	APPL	...

...	TABLEP	IDSETP	CTRL

Columns VARi represent rule variables values. The rest of the columns will now be explained.

As shown in the rules in the equation 2, the same predicate can be expanded more than once. For each expansion, we could use a different table PREDi_RULEj but it takes up a lot of space. Another option could be to use the same table PREDi_RULEj but we would have to distinguish between one and another expansion value set. If an order in the value sets is established, for instance the order in which they are introduced in the table, and this order is stored in the column IDSET, we can distinguish between several value sets. Furthermore, if we introduce special tuples in the table to separate the last value set of one expansion from the first set of the next expansion, the value sets are separated. The type of tuples is stored in the column CTRL and can take values 1 if the tuple is a value set one, or 2 if a separator. But the value sets in the same expansion have common data as do the variables of the value set that activated this expansion in the predicate, because these will have to be fixed when promoting values from the deeper predicate expansion. This information is common to all value set in the same expansion so the separator can be used to store it.

The mechanism can only process one value set simultaneously, so we have to provide a mark of the active value set. Once this is known, and knowing that the mechanism explores a tree of possibilities, the order in which it does this is not important. So, assuming a *stack* structure in the tables PREDi_RULEj allows the representation of the active set without the addition of a new column to store the active set mark. With this assumption, the last set is active.

The mechanism takes and treats a tuple each time and then gives control to another one. But we have to consider that logical rules allow recursiveness, that is, the rules defining an intensional predicate can contain the same predicate, so control can be transferred from a tuple corresponding to one expansion of the same predicate to another tuple corresponding to another expansion of the same predicate. Because of that, it is necessary to store however many predicates have been expanded for each tuple. The column dedicated to this is APPL. In addition, the number of fixed-value variables is stored in the column INST. Both two values are important because the difference between the two expansions of the same predicate is the number of fixed-value variables. A deeper expansion in the tree has a number of fixed-value variables greater than a superficial one.

When an active set has been filled in, the next step will be to promote variable values to the variables set which the expansion called predicate expansion. So the structure must be suited to store the identifier of the variable value set that called to the actual predicate expansion and the table

PREDi_RULEj containing it because the call can be made by the same rule of the predicate or another. This information is stored in the columns TABLEP and IDSETP.

The column NEG is represented if the value set satisfies the rule or can do it at the end of the expansion. It is useful when a negated predicate is being expanded because in this case the mechanism has to store the value set that satisfies or not the rule temporarily. It is possible that a value set satisfying the rule in the middle expansion (the value set has satisfied all predicate that has been expanded at the moment) can become into another not satisfying the rule at the end because it does not satisfy some of its predicates.

Comparisons must be considered in a special way. A comparison can occur in any position in the rule, but when it has to be expanded it is possible that the comparison variable is not fixed with a value, so the comparison cannot be expanded. If this case occurs then the comparison can be postponed to be expanded when the conditions for it are satisfied.

5.2 Deduction algorithm overview

When we expand a predicate, it can result in more than one values combination so we make a copy from the value set by expanding the predicate for every combination and each combination of values is used to fix a copy of the value set. Of course, the original value set is now no longer valid because there are actualized copies of it. Value 0 in the column CTRL is used to mark an invalid value set.

The mechanism can only obtain data when an extensional predicate is expanded resulting in a stack table called *PREDi* filled with the data from the extensional table. This is a step towards the unification of the mechanism. So there are two types of stack tables: those related to intensional predicates (PREDi_RULEj) and those related to extensional ones(PREDi).

The mechanism begins with one table PREDi_RULEj for each rule of the predicate expanded and an empty value set in each one of them. In the same way as only one value set can be active at a time, so can only one *stack table*, as tables PREDi_RULEj will be called from now on. So when it is necessary to expand another predicate, the mechanism has only to change the active stack table.

Once the stack tables for the expanded predicate have been initialized, the rule stack table is first marked as an active stack table. From now on, the process is common to any stack table.

The mechanism extracts the active value set from the active stack table. If it is a previously processed one then possibly the mechanism tried to expand a comparison that could not be expanded, so it was postponed. If the value set has postponed comparisons, the first step is to evaluate them so if they are not satisfied using the actual set of values, the value set is marked to be deleted. If the comparisons cannot be expanded then they are postponed again. Otherwise the value set satisfies the postponed comparisons, they are deleted and the mechanism continues with the processing of the active value set as specified in the following steps:

1. if the value set is not valid (CTRL=0), then it is modified (there are actualized copies of it) and the set and its postponed comparisons must be deleted.
2. if the value set is a valid one (CTRL=1), then there are predicates that have not been applied (the number of predicates of the rule minus APPL attribute is not zero), the next predicate is applied.
   (a) If the active stack table is intensional and has not been previously calculated, the mechanism determines the next predicate to be applied and its type and negation. Depending on the predicate type one of the following steps is performed:
      i. extensional predicate: the mechanism inserts an empty value set that is modified to ensure that the variables corresponding to predicate parameters and requiring to be fixed because of the call is made with fixed values. This modified set is inserted into the corresponding stack table PREDi_RULEj.
      ii. intensional predicate: the same method as that applied in the case of extensional predicate is carried out, but a modified empty value set is inserted into all the stack tables associated to the intensional predicate rules.
      iii. comparison predicate: the predicate is evaluated. If this is not possible, then comparison is postponed for this set. Otherwise, then the set must be marked as modified if it does not satisfy the comparison (CTRL = 0).

   The set APPL attribute is increased because a new predicate has been expanded. If the predicate is negated, the active set is marked as the set that activated negation. Finally, the next predicate stack table is set to active.

   (b) If the active stack table is previously calculated as extensional or intensional, the mechanism increases the APPL attribute in the active value set (APPL = 1), then it obtains data from the extensional or intensional pre-calculated table and inserts a copy of the active value set with the values modified for each tuple in the query result. The active value set is marked to modified (CTRL = 0).

   If there are no predicates to be applied two cases may occur:

   (a) If there are no fixed-value variables then no values combination satisfies the rule when all its predicates have been expanded. So the set can be deleted.
   (b) If there are no free variables, the separator for this expansion is used to promote the active set values to a copy of the value set that expanded the predicate. And then the caller set is marked to modified (CTRL = 0). The active set is deleted.

3. If the value set is a separator (CTRL=2), then all these expansion sets have been processed and we can mark to active the set that made this expansion.

Finally, if the value set is modified (so there are actualized copies of this, that is, CTRL=0), the set and its postponed comparisons are deleted.

In the case it is not modified and the active stack table is extensional (PREDi), then the data are obtained from the extensional table related to this and a copy for each combination of the result is introduced in the stack table. These data are those which satisfy the extensional predicate (i.e. those which are in the table, if the extensional predicate is not negated, or those which do not satisfy) are all of the possible data minus those which satisfy the predicate. The value set that expands the predicate is marked to modified. A predicate has been applied to all copies in the extensional stack table so the APPL increases by a unit.

6 Handling fuzzy information: fuzzy SELECT

The SELECT sentence is modified so that it can process flexible queries. There are the following modifications:

- Linguistic labels that start with the symbol '$'. As seen before, there are two types of labels: labels applied to ordered underlying domain attributes and those to non-ordered underlying domain attributes.
- Fuzzy comparators can either be possibility or necessity comparators. Possibility comparators are more generic and if applied, they return more tuples than necessity ones. All possible fuzzy comparators are shown in table 5. Some examples of fuzzy conditions with fuzzy comparators are the following:

```
Age FGEQ $Young
Sales_no NFLEQ $High
```

- THOLD<t> with $t \in [0, 1]$: it is a function specifying that the condition has to be satisfied with the minimum degree t. An example of the clause used is:

```
SELECT * FROM People WHERE Hair FEQ $Blond THOLD 0.5 AND
                           Height FGT $Tall THOLD 0.8;
```

- CDEG(<attribute_name>): it is a function that can appear in the select columns list. If the function appears in the select columns list, then a new column representing the compatibility degree of the attribute is added to the resulting table.
- CDEG(*): it is a function that can appear in the select columns list. If this call to function appears, then a new column representing the degree to which a tuple satisfies a condition is added to the resulting table.

- % character: this character is similar to the * character in the select columns list that makes all the table columns appear in the resulting table. But if the character % is used, then the compatibility degree of the relevant attributes is shown. These attributes are those appearing in the WHERE clause.
- Fuzzy constants: these constants are defined in section 4.1 and shown in the following table:

Fuzzy Const.	Syntax
UNKNOWN	UNKNOWN
UNDEFINED	UNDEFINED
NULL	NULL
Trapezoid-shaped pos. dist.	$\$[\alpha, \beta, \gamma, \delta]$
Linguistic label	$label
Interval	$[n, m]$
Approximate value	$\#n$

- IS condition: it is modified to accept fuzzy constants UNDEFINED and UNKNOWN.

```
<fuzzy attribute> IS [NOT] (UNKNOWN | UNDEFINED |
   NULL)
```

  Note that the use of the IS condition with a fuzzy constant and the use of the FEQ comparator with that constant does not have the same effect. The latter determines the compatibility degree or fuzzy attribute value with the constant.
- Fuzzy quantifiers: fuzzy absolute or relative quantifiers can be represented and used in the FSQL. The syntax to define them is:

```
$Cuantificador FUZZY[r] '(' condicion_difusa')' THOLD t
$Cuantificador FUZZY[r] '(' condicion_difusa1 ')' ARE
   (condicion_difusa2) THOLD t
```

7 Integrated fuzzy-deductive module

Deduction is a process that acts on an expansion tree. The middle nodes of this tree represent the values combinations that have not been completely instantiated and the leaf nodes those which have been completely instantiated. When we pass from a middle node to one which is below, we only instantiate variables. If the values promoted are fuzzy ones, then we also promote their compatibility degrees. So, fuzzy select is not applied at this point. The same occurs when we pass from a leaf node to a middle one. But when we are resolving a leaf node, i.e. an extensional or intensional previously calculated predicate, we must apply a fuzzy select that will return values and a

Type	Fuzzy Comp.	Meaning
Possibility	FEQ	Fuzzy Equal (on scalars or fuzzy numbers)
	FGT	Fuzzy Greater Than
	FGEQ	Fuzzy Greater or EQual
	FLT	Fuzzy Less Than
	FLEQ	Fuzzy Less or EQual
	MGT	Much Greater Than
	MLT	Much Less Than
Necessity	NFEQ	Necessarily Fuzzy EQual
	NFGT	Necessarily Fuzzy Greater Than
	NFGEQ	Necessarily Fuzzy Greater or EQual
	NFLT	Necessarily Fuzzy Less Than
	NFLEQ	Necessarily Fuzzy Less or EQual
	NMGT	Necessarily Much Greater Than
	NMLT	Necessarily Much Less Than

Table 5. Fuzzy comparators

compatibility degree. So, the satisfaction degree in which the values combination satisfies the predicate will depend on individual compatibility degrees. To summarize, modifications have to be made on the stack tables to store compatibility degree for the variables that can take fuzzy values and the rule satisfaction degree. The mechanism will have to be modified to manage these degrees.

8 Conclusions and future work

This work shows how the relational model can be extended to accept the representation and management of fuzzy and/or deductive information. This part is an extension of works that separately manage fuzzy information [7] or deductive information [16].

It works on the problem of creating a concrete implementation by means of the use of resources offered by a classical RDBMS like Oracle©.

The execution of a fuzzy query is only a translation matter but the execution of a deductive query implies a call to a procedure that has to execute several rules defining predicates involved in such a query. This procedure incorporates recursiveness management capabilities and it can not be implemented using relational languages so we have to turn to programming languages. Due to the practical application of the problem on the Oracle©RDBMS, there are two languages provided by the product that are the most interesting among all the possibilities: Pro*C©and PL/SQL©. Advantages and drawbacks of using one or another language are analyzed.

Finally, a built-in solution is chosen since it offers more advantages in data transfer between the DBMS and the inference engine. If we consider the given implementation, the language SQL has been improved so that it can accept new data types introduced (DDL) that can be fuzzy or intensional, and it can manipulate such data types (DML) (at the moment, only those sentences that are necessary for the basic execution of the deductive module have been implemented).

Two works would be interesting for immediate future. The first one is the possibility of using uncertainty in the rules for deduction, so the inference engine implemented has to be modified in order to add the features of the propagation model chosen. The second one is the implementation of remaining deductive and fuzzy DDL and DML sentences.

References

1. E.F. Codd. A relational model of data for large shared data banks. *Communications of the ACM*, 1970.
2. P. Bosc, M. Galibourg, and G. Hamon. Fuzzy querying with sql: Extensions and implementation aspects. *Fuzzy Sets and Systems*, 28:333–349, 1988.
3. B. P. Buckles and F. E. Petry. A fuzzy representation of data for relational databases. *Fuzzy Sets and Systems*, (7):213–226, 1982.
4. H. Prade and C. Testemale. Generalizing database relational algebra for the treatment of incomplete/uncertain information and vague queries. *Information Sciences*, (34):113–143, 1984.
5. M. A. Vila, J. C. Cubero, J. M. Medina, and O. Pons. Logic and fuzzy relational databases: A new language and a new definition. In P. Bosc and J. Kacprzyk, editors, *In Fuzzy Sets and Possibility Theory in Databases Management Systems*. Physica-Verlag, 1995.
6. M. Zemankova and A. Kandel. *Fuzzy Relational Databases - A Key to Expert Systems*. Verlag TUV Rheinland, 1984.
7. J.M. Medina, O. Pons, and M.A. Vila. Gefred. a generalized model of fuzzy relational databases. *Information Sciences*, 1994.
8. J.M. Medina, O. Pons, J.C. Cubero, and M.A. Vila. Freddi: A fuzzy relational deductive database interface. *International Journal of Intelligent Systems*, 12:597–613, 1997.
9. L.A. Zadeh. The concept of a linguistic variable and its application to approximate reasoning. *Information Sci.*, 8, 9:(8) 199–248, 301–357, (9) 43–80, 1975.
10. J. Grant. Incomplete information in a relational database. *Fundamenta Informaticae*, 3:363–378, 1980.
11. E. F. Codd. Extending the database relational model to capture more meaning. *ACM Transactions on Database Systems*, 4:262–296, 1979.
12. E. F. Codd. Missing information (applicable and inapplicable) in relational databases. *ACM SIGMOD Record*, 15(4), 1986.
13. E. F. Codd. The twelve rules for relational dbms. Technical Report EFC-6, San Jose, The Relational Institute, 1986.
14. E. F. Codd. More commentary on missing information in relational databases. *ACM SIGMOD Record*, 16(1), 1987.

15. J. Galindo. *Tratamiento de la Imprecisión en Bases de Datos Relacionales: Extensión del modelo y adaptación de los SGBD actuales.* PhD thesis, Department of Computer Science and Artificial Intelligence, University of Granada, SPAIN, 1999.
16. O. Pons, J. M. Medina, J. C. Cubero, and A. Vila. *Foundations of Intelligent Systems*, chapter An Architecture for a Deductive Fuzzy Relational Database. Springer, 1996.

A Fuzzy SQL sentences syntax

A.1 FSQL DDL sentences

All fuzzy catalog tables are managed using specific sentences created on the basis of classical SQL sentences. We will present sentences in EBNF notation but we will explain only those items that are not typical.

New or modified sentences are the following:

Create table : a classical SQL sentence is improved to accept fuzzy types as column type.

```
create_table    : CREATE TABLE table_item '(' column_list ')'
    classical_create_table_clauses ';'
column_list     : column_list ',' column_list | column_descpt |
    table_restriction
column_descpt  : column_id datatype
datatype        : classical_datatype [classical_restriction_list]
   | fuzzy_datatype [fuzzy_restriction_list]
fuzzy_datatype : fuzzy_type1 | fuzzy_type2 | fuzzy_type3
fuzzy_type1     : (FTYPE1 | CRISP) '(' margin ',' much ')'
   [base_type]
fuzzy_type2     : (FTYPE2 | POSSIBILISTIC) '(' margin ',' much
   ')' [base_type]
fuzzy_type3     : (FTYPE3 | SCALAR) [ '(' len ')' ]
   [DOMAIN column_reference]
fuzzy_restriction_list : NULL | ONLY LABEL | NOT not_allowed
not_allowed     : NULL | UNDEFINED | UNKNOWN | LABEL | CRISP |
   TRAPEZOID | INTERVAL | APPROX
base_type       : classical_datatype
```

margin and *much* are parameters associated to fuzzy type 1 and 2 attributes and *len* is another one associated to fuzzy type 3. Fuzzy type 3 attributes admit compatibility relations so all labels and distributions owned by the fuzzy database column *column_reference* are copied to be owned by this if the *DOMAIN* clause is present in the sentence. *classical_datatype* refers to typical data types in the RDBMS. *classical_restriction_list* is a list of typical RDBMS column restrictions. *classical_create_table_clauses* refers to typical clauses appearing in the classical CREATE TABLE sentence (STORAGE, CLUSTER, ...).

CREATE LABEL : new sentence that creates linguistic labels for fuzzy type 1 and 2 attributes.

```
create_label       : CREATE LABEL create_label_type ';'
create_label_type  : create_label_type1 | create_label_type2
```

```
create_label_type1 : ID ON column_reference VALUES alpha ','
   beta ',' gamma ',' delta
create_label_type2 : '*' ON column_reference FROM
   column_reference
```

Two ways of creating labels exist. The first one (*create_label_type1*) is an explicit creation, and the second one (*create_label_type2*) is an implicit creation by copying labels associated to another fuzzy type 1 or 2 attribute. *column_reference* attribute must be a fuzzy type one.

CREATE NEARNESS : a new sentence that creates linguistic labels for a fuzzy type 3 attribute and similarity values for every possible values pair.

```
create_nearness : CREATE NEARNESS ON column_reference LABELS
   label ',' label {',' label}* VALUES similarity_value
   {',' similarity_value}*
```

Note that a minimum of two labels and one similarity value is required. If n is the number of labels then it is necessary to define the number of similarity values:

$$\text{similarity values number} = \frac{(n * n)}{2} - \frac{n}{2} \tag{7}$$

An example of this sentence use is:

```
CREATE NEARNESS ON a
LABEL b1, b2, b3, b5, b5, ..., bn
VALUES b1#b2, b1#b3, b1#b4, b1#b5, ..., b1#bn,
              b2#b3, b2#b4, b2#b5, ..., b2#bn,
                     b3#b4, b3#b5, ..., b3#bn,
                            b4#b5, ..., b4#bn,
                                   ...
                                        b(n-1)#bn;
```

ALTER TABLE : the classical sentence is modified in clauses ADD and MODIFY:

```
alter_table : ALTER TABLE table_name change {',' change}
   alter_tab2
change      : ADD '(' column_list ')' |
              MODIFY '(' column_descpt {',' column_descpt}* ')'
```

column_list and *column_descpt* have been previously defined and the rest of the sentence clauses remain the same as in the classical sentences.

ALTER LABEL : is the same as that used to execute a *DROP LABEL* sentence followed by an *ALTER TABLE*.

ALTER NEARNESS : a new sentence that allows the addition of new labels and associated similarity values or the modification of a similarity value between two existent labels.

```
alter_nearness : ALTER NEARNESS ON column_reference alter_clause
alter_clause   : alter_clause1 | alter_clause2
alter_clause1  : LABELS label, label {',' label} VALUES
   similarity_value {',' similarity_value}
alter_clause2  : CDEG '(' label ',' label ')' '='
   similarity_value
```

DROP LABEL : a new sentence that destroys labels associated to fuzzy type 1 and 2 attributes.

```
drop_label : DROP LABEL (* | label) ON column_reference
```

label is only used with fuzzy type 1 attributes.

DROP NEARNESS : a new sentence that destroys the labels associated to fuzzy type 3 attributes and deletes their similarity values.

```
drop_nearness : DROP NEARNESS [label] ON column_reference
```

If *label* appears in a sentence, only the label with the identifier *label* will be destroyed and its similarity values will be deleted.

A.2 FSQL DML sentences

INSERT INTO : this sentence remains the same but new value types appearing in the values list are added.

SELECT : the classical sentence accepts new value clauses to represent fuzzy constants in the *where* clause. The functions set is also increased to add membership degree functions and fuzzy comparison operators.

DELETE FROM : this sentence has a *where* clause so considerations for the *SELECT* sentence are applied to this.

B Deductive SQL sentences syntax

B.1 DSQL DDL sentences

Due to a new table type and the fact that new data structures are required, a whole new syntax is created.

CREATE INTENSIONAL TABLE : this sentence creates a typical table but associates its procedures and triggers which allows the maintenance of an actualized database.

```
create_intensional_table     : CREATE INTENSIONAL TABLE
   table_name '(' intensional_columns_list ')' ';'
intensional_columns_list  : intensional_column_descpt
   {',' intensional_column_descpt}*
intensional_column_descpt : column_name classical_datatype
```

classical_datatype refers to implemented RDBMS data types. This sentence modifies

CREATE RULE : once the intensional table has been created, it is necessary to associate the rules that the inference engine explores to complete the table. These rules are defined by this new sentence with the following syntax:

```
create_rule     : CREATE RULE FOR intensional_table_name '('
   variables_list ')' AS predicates_list ';'
variables_list  : identifier ',' identifier {',' identifier}*
predicates_list : predicate {AND predicate}*
predicate       : [NOT] (intensional_predicate |
   extensional_predicate | comparison
intensional_predicate : predicate_name '(' variables_list ')'
extensional_predicate : predicate_name '('
   extensional_predicate_variables_list ')'
extensional_predicate_variables_list :
   extensional_predicate_variable {','
   extensional_predicate_variable}*
extensional_predicate_variable : identifier ':'
   table_column_reference
comparison      : ('=' | '<''>' | '>''=' | '>' | '<''=' | '<')
   '(' identifier ',' identifier ')'
```

variables_list is the list of variables that will appear in the rule body instantiated in the predicates. As has been previously shown and stated, the rules are conjunctive. The predicates can be intensional or extensional. When the called predicate is intensional, the variables specified will be instantiated by its rules expansion. When the called predicate is extensional, the variables must be associated to extensional table columns so that the deduction algorithm can obtain data for this predicate.

Some restrictions imposed on the definition of rules are:

- It is not possible to have no jointed variables, i.e. variables which only appear once in the rule. These are variables that never instantiate and are not allowed. item rule body can contain intensional or extensional predicates and comparisons. The predicate defined in the head rule can appear in the rule body so the rule becomes a *recursive rule.*
- Variables must be jointed, i.e. they appear more than once in the rule. When they are a comparison, variables must appear in another predicate not being comparison.
- The first predicate not being comparison in the rule body cannot be the predicate in the rule head, because of the infinite loop of recursiveness. This problem will be explained later.

DELETE RULE : created rules must can be deleted and this sentence makes it possible using the following syntax:

```
delete_rule : DELETE RULE ('*'|number) FOR
   intensional_table_name ';'
```

It allows the deletion of one specific rule identified by *number* or all the rules associated to an intensional table using *.

DROP INTENSIONAL TABLE : this new sentence drops an intensional table created previously.

```
drop_intensional_table : DROP INTENSIONAL TABLE
   intensional_table_name
```

The intensional predicate associated to the table cannot be referred by other rules. If it is, the drop will fail.

B.2 DSQL DML sentences

SELECT sentence : the classical sentence is not modified in syntax but it is in management. When a table name appears in FROM section, the system will explore to see if it is intensional or extensional. If it is extensional, there is nothing to be done. If it is intensional, the system will use the rules defining the table to complete it.

The Fuzzy Object-Oriented Database Management System

Gloria Bordogna[*], Alberto Leporati [°], Dario Lucarella[+], Gabriella Pasi[*]

[*] Institute for Multimedia Information Technologies, National Research Council, Via Ampère 56, 20131 Milano, Italy

[°] Department of Computer Science, University of Milan, Via Comelico 39, 20135 Milano, Italy

[+] Automatica Research Center, National Electrical Energy Company, Via Volta 1, Cologno Monzese, (MI), Italy

Abstract. In this paper we present a prototypal implementation of the Fuzzy Object Oriented Data model[1]. FOOD has been defined for representing data characterized either by imprecision or vagueness and uncertainty. The present implementation extend the O2 system so as to allow the representation and instantiation of vague attribute values and uncertain and strengthened relations.

Keywords. Object Oriented Database models; possibility theory; fuzzy set theory.

1 Introduction

In this paper the design of a prototypal DBMS implementing the Fuzzy Object Oriented Data model (FOOD) is presented. The FOOD model [1,2] has been designed to manage imperfect information, i.e. imprecise or vague and uncertain information [3], both at the level of data representation and in queries, and has been formalized within the framework of fuzzy set and possibility theory [4,5]. By means of the FOOD model it is possible either to deal with a variety of data richer than those traditionally managed by crisp Object Oriented database models as well as to soften the interaction with the user. In fact, the FOOD model supports a graph-based representation [6,7]: both the database scheme and instances are represented as directed labeled graphs, and users can manipulate and retrieve information just by directly performing actions on the graph of the conceptual schema visualized on the screen without submitting requests in a traditional query language [8]. Distinct graphical elements are used to represent the different kind of imperfection of the information so that the user is made aware

[1] This work is partially supported by the National Electrical Energy Company (ENEL) under the Research Grant RAUWC0005.

of the fact that the database does not always contain a precise description of the real world entities, rather it specifies a set of constraints which the actual entities satisfy to some extent: such a characteristics makes the FOOD model particularly suitable to be integrated with a representation based query languages, such as the one proposed by Bosc and Pivert in [9].

The present implementation of the FOOD model, that we call FOOD system, is a prototypal one, realized to the purpose of testing the model on real data. It has been designed as an extension of the commercial product O_2 – by O_2Technology [10]: all the feaures of O_2 system are then inherited by the FOOD system thus relieving the developer from taking care of the management of processes typical of an OODBM environment.

Particular care has been devoted to design a flexible implementation, easily modifiable with future extensions. The fuzzy pattern matching mechanism [11], necessary for implementing the evaluation of queries against vague data, has not been implemented yet, but the data structure storing vague attribute values has been designed so as to simplify its management by a future implementation of the retrieval mechanism.

Actually, the FOOD system allows both the creation of a FOOD conceptual schema and the navigation into the database through a graphical interaction. The user is faced with a graphic environment, in the style of the *Msdraw* package, and can build the conceptual schema of the database by drawning on a virtual canvas the graphic elements representing the modeled entities and their interrelationships; then, the conceptual schema can be populated by clicking with the mouse on the desired graphic entities visualized onto the canvas. Also the visualization of the database content is supported by a graphical interaction on the conceptual schema. However, the database instances are not visualized in a graphical form as prescribed by the FOOD model, since from a practical point of view, this modality did not seem to further improve the user system interaction; rather we estimated that a graphic representation of the instances would create confusion in the case the retrieved entities have a complex structure, due to the great number of highly interconnected elements laid one upon the other visualised on the screen.

In the next section the formalization of the FOOD model is introduced; in section 3 the different kind of imperfect information that can be represented in the FOOD model are illustred. In section 4 the architecture of the FOOD system is described; in section 5 the O2FOOD schema is presented, and in section 6 the functional characteristic of the user interface are described. Finally in section 7 concluding remarks and future developments are discussed.

2 The FOOD model

FOOD is a Fuzzy Object-Oriented Data model defined as an extension of a Graph-based object model in order to manage both crisp and imperfect information [1,2]. The need for representing information affected by vagueness and/or uncertainty is a requisite of many current applications dealing with data of different nature and with complex interrelationships. The FOOD model is formalized within fuzzy set theory [4] and possibility theory [5] and is based on a visual paradigm by means of which both the database schema and instances are represented as directed labeled graphs [6,7]. Here, we briefly recall the formalization of the FOOD scheme and of the FOOD database instances.

The conceptual scheme is defined as a quintuple: {C, T, A, H, P} in which:

C $= C_c \cup C_f$ is a finite set of class names in which C_c is the set of names of crisp classes and C_f is the set of names of fuzzy classes; at the intensional level each class $c \in C$ denotes a structure, i.e., a set of attributes and a behavior; at the extensional level it identifies the collection of objects having that structure. More precisely a crisp class collects objects with a full membership while a fuzzy class collects objects which can have a partial membership to the class.

T is a finite set of type names; each $t \in T$ denotes a type collecting primitive objects. Types can be either crisp T_c or vague T_V depending on the fact that there elements denote sets of either precise values or vague and imprecise values.

A is a set of attribute names; there are two ways to classify the elements in A: attributes are simple when their domain is a type $t \in T$, they are complex when their domain is a class $c \in C$. Moreover, it can be made a distinction between single-valued attributes A^s (they take a single element of the domain as value) and multi-valued attributes A^m (they take a set of elements of the domain as value), with $A = A^s \cup A^m$. Both A^s and A^m may be simple or complex.

P $\subseteq C \times A \times (C \cup T)$ is the property relation. If $(c_i, a, c_j) \in P$, then class c_i has the attribute named a, whose domain is the class c_j. If $(c_i, a, t_j) \in P$, then class c_i has the attribute named a, whose domain is the type t_j. A property relation can be strengthened P_S when the attribute values are associated with an object by specifying an intensity or it can be uncertain P_U when the associated attribute values are not the only possible values. Finally it can be both strengthened and uncertain P_{SU}.

H $\subseteq (C \times C) \cup (C_f \times C_f \times Labels)$ is the inheritance relation that allows to define class hierarchies;

Labels = {*not completely, more or less, very, definitely* }

is the set of linguistic modifier names; a modifier in *Labels* is used in the fuzzy inheritance relation to specify the extent of the generalization or specialization; a class c_i that is defined as subclass of a class c_j inherits the attributes of c_j, and the objects (or instances) of the subclass are also objects of the superclass. If the inheritance relation is fuzzy the value $m \in Labels$ associated with a subclass-superclass pair $(c_i\ c_j)$ specifies the extent to which the instances of the subclass c_i are also instances of the superclass c_j

The conceptual scheme can be represented as a directed labeled graph $G(\Sigma) = (N, E)$ in which:

N = (C $\cup$ T) is the set of nodes; each class $c \in C$ is represented by a rectangular-shaped node labeled c. Each type $t \in T$ is represented by an oval-shaped node labeled t. Fuzzy classes are represented by rectangular shaped nodes filled with a shaded pattern. Vague types are represented by oval-shaped nodes filled with a shaded pattern.

E = (H $\cup$ P) is the set of edges. For each $h_{ij} \in H$ there is associated an edge labeled "*is a*" directed from class c_i to class c_j. If the inheritance relation is fuzzy the corresponding edge is in bold and is labeled "*is a m*" with *m*$\in$*Labels.* For each (c_i , a , c_j) $\in$ P there is an associated edge labeled a, directed from c_i to c_j. In particular, if $a \in A^s$ the edge has a single arrow, while if $a \in A^m$ the edge has a double arrow. Further edges corresponding to strengthened relations are in bold. In Figure 1 the graph representing the FOOD conceptual schema describing research projects is shown.

The database instances over a scheme are defined by the quintuple S = (Σ, O, V(P), I, L) in which:

Σ is the conceptual scheme defined previously;

O is the set of actual objects stored into the system;

V(P) is the set of actual attribute values stored into the system, in which $P \subseteq C \times A \times T$; $\forall p= (c,a,t) \in P$ with $c \in C$, $a \in A$ and $t \in T$, there exists an attribute value Val(p)=Val(c,a,t) $\in$ V(p); in Figure 2 examples of elements in V(P) set are given by Val(person, name, string) = {M.Rossi, V.Massi}, Val(project, name, string) = {FIRST}, Val(product, name, string) = {FOOD prototype};

I $\subseteq (O \times C_c) \cup (V(P) \times T) \cup (O \times C_f \times$ *T(Strength)*) is the instance relation; each object $o \in O$ is an instance of a crisp class $c \in C$, and each value Val(c,a,t)$\in$ V(P) is an instance of the type $t \in T$; further if $c \in C_f$, i.e. c is a fuzzy class, the object is an instance of the class with an intensity specified by a linguistic value chosen in the term set of the variable *Strength*:

T(*Strength*)={*none, very low, low, medium, high, very high, full*} (1).

An ordering is defined between the values of the term set *T(Strength)* so that a value in it corresponds to a higher (lower) strength than its preceding (following) value; *none* corresponds to no strength, *full* corresponds to maximum strength.

L $\subseteq$ O $\times$ A $\times$ (**P**(O [$\times$ *T(Strength)*]) $\cup$ **P**(V(P) [$\times$ *T(Strength)*])) [$\cup$ [0,1]] is the link relation in which **P**(.) denotes the power set and the square brackets delimit the "optional" elements. *T(Strength)* is mandatory for strengthened properties, i.e. when an object is associated with another object or attribute value by specifying an intensity chosen in the term set *T(Strength)*. The set [0,1] is manatory for uncertain properties, i.e. when an object is associated with an attribute value by specifying a degree of uncertainty in [0,1]. In the instance graph depicted in Figure 2, an example of simple link relation is (O_{pub}, due date, June 2000) in which O_{pub} refers to the paper titled "A fuzzy object-oriented data model"; and an example of complex link relation is (O_{proj}, leader, O_{person}), in which O_{proj} refers to the project named "FIRST", and O_{person} refers to Mr."M.Rossi".

Given the scheme graph G(Σ), the instance graphs can be generated through the application of the instance relationship; in an instance graph the nodes correspond to actual objects and values stored into the system.

3 Imperfect data representation in FOOD

The imperfection in data representation is modeled in FOOD in order to define:
1. vague attribute values;
2. uncertain property and uncertain link relationships;
3. strengthened property and strengthened link relationships;
4. fuzzy classes;
5. fuzzy class hierarchies.

In the following we synthetically analyze the formalization of the previous five entities in FOOD.

3.1 Vague attribute values

Vague attribute values are defined to model situations in which the precise value of a given object attribute is not known precisely; a vague value for the attribute, is in this case specified as a linguistic value associated with a possibility distribution over the attribute domain. Formally, a simple attribute *a* of a class *c* that may take vague values is associated with a linguistic variable named *c.a*. The attribute domain D is in this way extended with the term set of the linguistic variable T(*c.a*) referred to as a vague type T_V, so as to become $D_e = \mathcal{P}(D) \cup T(c.a)$, in which $\mathcal{P}$ denotes the power set. Vague attribute values are then linguistic terms drawn from the term set of a linguistic variable. For example, the domain of the crisp attribute age is the set of integers D = [0..120]∈T. The vague type T(*person.age*)∈T_V corresponds to the term set of the linguistic variable named *person.age*. The linguistic values of *person.age* are for example *old, young*∈T(*person.age*) defined as possibility distributions on the base variable domain D=[0..120]. The domain of the attribute *age* of a *person* is then extended to take values in D_e= [0...120]∪ T(*person.age*). For example, the scheme graph in Figure 1 has vague types: T(*person.age*), T(*person.expertise*), T(*product.price*), T(*realization date*) ∈ T_V which are defined to extend the domains of the attributes "age", "expertise", "price" and "date" respectively. Their values are expressed as possibility distributions on the basic domains "integer", "string", "real" and "date" respectively. In the example, the possible vague values can be expressed by means of linguistic terms belonging to T(*person.age*) = {*very young, young, medium aged, old, very old*}, T(*person.expertise*) = {*researcher, expert, very expert, project manager*}, T(*product.price*) = {*very low, low, high, very high*}, T(*realization* date) = {*not recent, not very recent, recent, very recent*} respectively.

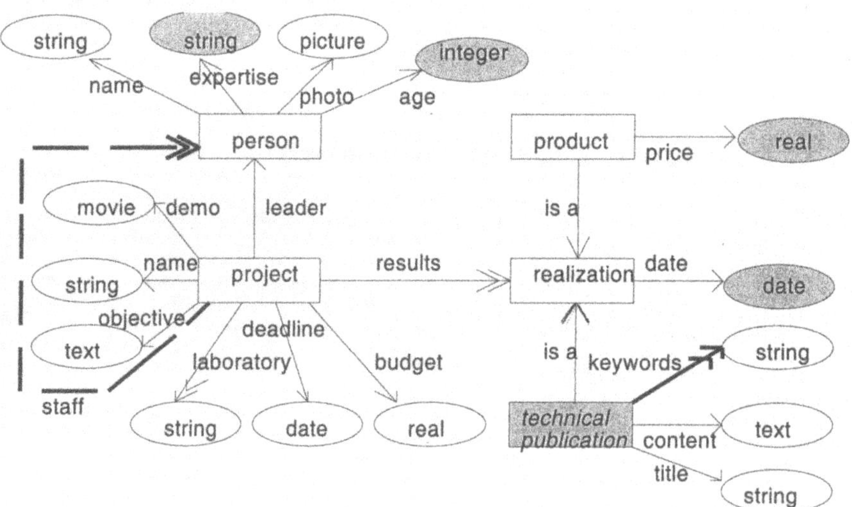

Figure 1: FOOD conceptual schema of a database describing research projects

3.2 Uncertain properties of objects

The lack of complete information on the actual value of an attribute can lead to associate a degree of uncertainty with the selected attribute value. For example, let us suppose that one is not sure that the "budget" of the "project" is "80.000 Ecu". This situation can be represented by associating an uncertainty degree $\varepsilon \in [0,1]$ with the link associating the "project" named "*FIRST*" with the "budget" value "80.000Ecu". The uncertainty degree ε can be interpreted as a tolerance threshold to violate the constraint imposed on the actual attribute value by the specified attribute value. The higher the value ε, the higher is the tolerance to violate this constraint, i.e. the higher is the possibility that the actual attribute value is different from that specified "80.000Ecu". In the FOOD scheme, the uncertainty in associating a domain value with the attribute of an object can be formally represented through the definition of an uncertain property P_u. For each uncertain property relation $p \in P_u \subseteq P$ in the schema Σ there exists an uncertain link relation $l \in L_u \subseteq L$ in the corresponding database; in the case of a single valued attribute $a \in A^S$ the uncertain relation $l \in L_u$ associates with an object $o_k \in c_i$

one single value $o_h \in c_j$ with an uncertainty degree $\varepsilon \in [0,1]$. In case of multi-valued attributes $a \in A^m$ the uncertain relation $l \in L_u$ associates with an object $o_k \in c_i$ m values $o_1 \dots o_m \in c_j$ with an uncertainty degree $\varepsilon \in [0,1]$.

In the scheme graph in Figure 1 five uncertain property relations have been defined:

(product, price, P(real)∪*T(product.price)*), (realization, date, P(date)∪ *T(realization, date)*), (project, budget, real), (project, deadline, date) and (project, staff, person); it can be noticed that the domains can be either crisp types (real, date), or extended to include vague types (*T(product.price)*, *T(realization, date)*) or classes (person).

In Figure 2 an instance graph of the scheme graph in Figure 1 is shown. The project "FIRST" is known to have a "budget" of "80.000 Ecu" with an uncertainty equal to 0.3 (it is almost certain, but it cannot be excluded that the budget can have a different value when the project is funded). The "deadline" of the "project" is very uncertain with a degree equal to 0.8 (it indicates that the deadline is very likely to be different; this may suggest that it will be probably postponed). The "person" "V. Massi" is indicated as a member of the "staff" of the project with an uncertainty 0.1 (it is almost certain that this person will participate at the project but it cannot be excluded the participation of other persons). Further the "project" will produce a software "product" named "FOOD prototype" whose "date" is foreseen for "*spring 2000*" with an uncertainty equal to 0.4 (this suggests that the " date" is subject to changes) and whose "price" is "*around 3000 Ecu*" with an uncertainty 0.5 (the "price" value is approximate but there is still a tolerance for other values). Finally the "project" will yield a "paper" that will be published during the "*year 2000*" with an uncertainty degree equal to 0 (the "paper" has been accepted and is scheduled by the editor in journal for 2000).

At this point one could ask the way in which the uncertainty degree of the link relationship and the vagueness of the attribute values affect the evaluation of queries. For instance, a query asking for the "products" with a "price" higher than "1000 Ecu", is translated into the following selection condition:

"val(project,price, P(real)∪T(*product.price*))>1000 Ecu"

with (project,price, P(real)∪T(*product.price*))$\in P_u$.

This condition must evaluate values which may be vaguely or imprecisely defined, e.g. "*around 3000 Ecu*", and which are associated with a "project" with an uncertainty degree, e.g. (*oid*, price, *around 3000 Ecu*)=0.5 in which *oid* is the object identifier of the product named "FOOD prototype". To this aim the uncertainty degree ε of the link relation is used to modify the possibility distribution π representing the attribute value $\pi_{stored} = \max(\varepsilon, \pi)$ [12]. It must be noticed that this process can be performed once and for all when creating or updating a database. In the cases of both precise and imprecise attribute values, the same procedure is applied since also these values are represented by possibility distributions π. The evaluation of the query reduces then to the fuzzy pattern matching between the modified possibility distribution, in the example

π_{stored} =max(0.5,$\pi_{around\ 3000\ Ecu.}$), and the fuzzy set representing the desired attribute values specified by the query condition, in the example: $f_{desired\ values} = \int_{i=0}^{1000} 0/i + \int_{i=1000} 1/i$. Notice that this approach can be used even in case the query contains vague conditions, such as "*expensive*" [11, 12].

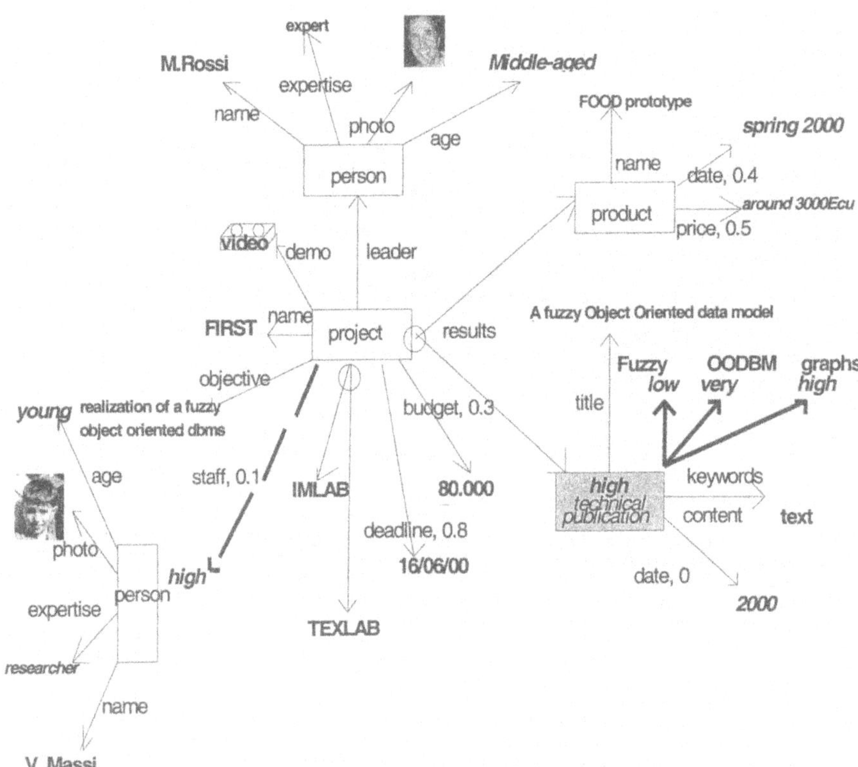

Figure 2: An instance Graph over the FOOD conceptual schema graph in Figure 1.

3.3 Strengthened properties of objects

Another useful modeling feature is the expression of the strength of the property relationship. It can be formally represented through the definition of a strengthened property P_s in the FOOD scheme. Also an uncertain property can be enriched with strength; correspondingly a set of properties P_{su} are defined in the FOOD scheme. In the scheme graph G(Σ), for each strengthened property relation $p \in P_s \subseteq P$ in the scheme, there exists a strengthened link relation $l \in L_s \subseteq L$ in the corresponding database, denoting that the attribute a of an object o_k instance of a

class c_i has the value o_h instance of a class c_j (or the m values o_1 ,..., $o_m \in c_j$) with a degree (m degrees) of strength expressed in the form of a linguistic label (m linguistic labels) belonging to the term set *T* of the linguistic variable *Strength* (see definition (1)).
In the scheme graph in Figure 1 two strengthened property relations are introduced "staff " and "keywords"; the first relation is also uncertain. In the corresponding instance graph in Figure 2 the strengthened link relations are shown: the strength of involvement of "Mr.V.Massi" in the "project FIRST" will be *high* conditioned to the fact that his participation to the project is still uncertain (0.1). Finally, the paper titled "A fuzzy Object-Oriented Data Model" deals with a different strength (having the semantics of importance) with the concepts represented by each of the "keywords": it deals *low* with "fuzzy", *very high* with "OODM" and *high* with "graphs". It is clear from the example that the semantics of the strength is strongly dependent on the semantics of the elements involved in the relation.

3.4. Fuzzy classes

There are real situations in which it is useful to be able to represent the partial membership of an object to a class. These cases often occur when the concepts defining the class are vaguely formulated, and the membership of an object to the class is dependent on the satisfaction of a fuzzy constraint imposed on some attribute values of the object.

The handling of fuzzy classification of objects is achieved in FOOD by defining fuzzy classes C_f, characterized by objects whose membership to the class is gradual. Since this gradualness expresses the strength of the instance relationship, it is represented in FOOD by a linguistic qualifier such as those already defined for the strength of the link relation (see definition (1)). A fuzzy classes is then characterized by the strengthened *instance* relations $i \in I_s$. A strengthened instance $(o_k, c_i, s) \in I_s$ associates the object o_k with the fuzzy class c_i, with a membership degree expressed by $s \in$ *T(Strength)*. In the schema graph in Figure 1 there is the fuzzy class *technical publications*. The "paper" titled "A Fuzzy Object-Oriented data model" in Figure 2 is an instance of the fuzzy class *technical publications* with a strength equal to *high.*

3.5 Fuzzy class hierarchies

In this context, the definition of fuzzy class hierarchies is considered a useful mean to represent the vagueness in a hierarchy defined for classification purposes. This situation occurs when both the superclass and the subclass are fuzzy: in fact, due to this fact the vague concepts represented in the superclass can be specialized or generalized by the vague concepts represented in the subclass to a vague extent, which can be specified by a linguistic modifier such as "*very*" for specialization and "*more or less*" for generalization. Notice that, the concept of class specialization implies the inclusion of the subclass in the superclass: an object that is an instance of a subclass with a membership strength equal to *m*, is also an instance of the superclass with a membership strength equal to *m1>m*. The contrary applies in case of generalization. To this aim, the fuzzy inheritance partial ordering relation H_f is defined as: $H_f \in (C_f \times C_f \times Labels)$, in which *Labels* is a set of operator names applying modifications of the membership strength of an object to a fuzzy class, e.g. *Labels={not completely, more or less, very, definitely }*. $(c_i , c_j , very) \in H_f$ indicates that the fuzzy class c_i specializes the concepts in the fuzzy superclass c_j to the extent *very*. Notice that in this context, these operators are not defined as linguistic modifiers (Their definition can be found in [2]).

4 The architecture of the FOOD prototypal system

The implementation of the FOOD model has been conceived as an extension of the commercial product O_2 – by O_2Technology – an Object Oriented Database Management System compliant with the ODMG-93 standard [13]. O_2 is programmable by means of a proprietary language – O_2C – which is an object oriented extension of the C language, with direct access to data types and constructs defined by the O_2 Data Definition Language; moreover, the elements of an O_2 database are easily accessible from programs written in C or C++ by means of a *binding* protocol, implemented as a collection of library functions. The O_2 System has a client/server architecture; this means that every operation on the schema or on an object associated with it is invoked by the user from a *client* program and executed by the *server*. The graphical interface that is part of the implementation of the FOOD model is a new client for O_2, by means of which the user can design his/her own conceptual schema in a visual way and populate it with objects.

As it has been explained in the introduction, the FOOD model is based upon a *visual* paradigm of interaction with the user: through a graphical interface, he/she can design the conceptual schema for a database by viewing it as a graph pictured onto a *canvas*. Every element of the schema is represented using distinct graphical symbols according to the definition of the model. When the user performs a graphical operation onto the graph (addition, removal or modification of a node or arrow), an appropriate sequence of commands is invoked to the O_2 System, which executes them in order to reflect the occurred changes into the database.

From a practical point of view, the user interface allows the user to design a given schema and populate it without even knowing the so-called *Data Definition Language*, because every operation is performed in a graphical way. The knowledge of the O_2C programming language is needed only if the user wants to add some methods to the classes present in his/her schema. Further it is possible to *navigate* through the set of instances of the schema without knowing the *Object Query Language* (OQL).

For implementing the user interface, a new client program has been written in C with the Motif graphics libraries; the graphics capabilities provided by the O_2 modules `o2look` and `o2graph` have not been used extensively due to their limited flexibility in the management of the graphical elements that characterize the FOOD model. The application that provides the user interface has the form of a standard Unix executable file which – by means of the O_2 binding libraries – connects to a given named system[2] containing both the schema under development and a base associated with it. Commands are invoked from the graphic interface by selecting the elements onto the canvas and/or pushing one *command button*; as a result, every operation on the graph representing the conceptual schema invokes an appropriate function in the O2FOOD schema (as it will be explained in section 3), also contained into the named system. Thus, the implementation is structured in two distinct but interconnected parts:

1) the O2FOOD schema, a collection of O_2C classes and functions that allow to deal with the entities in the model, and some operations on them;
2) the user interface (that is, the new client) that accepts the graphical commands from the user and calls the appropriate functions in the O2FOOD schema to create/remove new classes, attributes, crisp or vague types and objects.

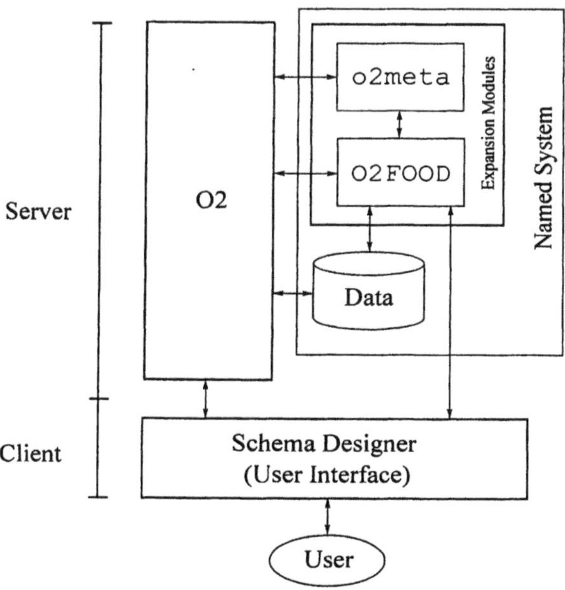

Figure 3: Relations occuring between the O2FOOD schema and the User Interface with respect to the elements of a typical O_2 application.

[2] In O_2 terminology, a *named system* is a collection of conceptual schemas – both user- and system-defined – plus the associated bases.

In Figure 3 we can see how the O2FOOD schema and the user interface relate to the other elements of a typical O_2 application. It is evident from the Figure how the O2FOOD schema can be considered a logical extension of the O_2 system, as happens with the modules already present in O_2 (`o2look`, `o2meta`, `o2kit`, etc.).

Because of the difficulties involved in programming an effective and user-friendly event-driven interface in Motif (the target windowing environment), we made a preliminary study of the operating features of the user interface with a Rapid Application Development environment – actually, Microsoft Visual Basic 5.0 with the help of an Access 97 database. This approach allowed us to experiment various solutions about the structure of the user interface without having to rewrite big portions of code. The implementation process has thus been carried out in the following three phases:

1) the definition and implementation of the O2FOOD schema;
2) the preliminary study of the user interface with Visual Basic 5.0 and Access 97;
3) the porting of the interface program from the PC to a SUN SparcStation 20 workstation, running Solaris Version 2.5 with Motif and the O_2 System Version 5.0.

In the following two sections we present the elements of the O2FOOD schema, and the use and implementation issues of the user interface. The pictures shown from now on are screen shots taken – for sake of simplicity – from the prototype interface written on the PC.

5 The O2FOOD schema

The O2FOOD schema can be thought as a logical extension of O_2 which enables to deal with imprecise, vague and uncertain information; this capability is achieved through the definition of the classes and data structures that allow to work with vague types (as defined by the FOOD model) over a wide range of basic domains, both discrete and continuous and with uncertain and strengthened property relations.

In Figure 4 the logical structure of the O2FOOD schema is sketched: as we can see, there is one class for every kind of basic domain we have considered. The structure of each class is designed to represent efficiently a possibility distribution over the corresponding domain, while the operations that can be performed on such possibility distributions (which represent both the vague type values and the uncertain property relations) are implemented as methods for the class.

For discrete basic domains, possibility distributions are represented by pairs (x, y) of values, where x is an element of the basic domain that can be an integer value, a real value or a string, and $y \in [0, 1]$ is the membership degree of x. The methods of the class implement the usual operations of intersection, union and complementation of fuzzy sets with a discrete integer domain (defined respectively by computing the minimum, maximum and 1's complement on the corresponding membership functions), plus other operations that will be useful in future implementations and extensions of the FOOD model, such as the horizontal translation of the membership function, the computation of the membership degree for a given value of the domain.

The continuous possibility distributions over the real domain are defined as *piecewise-linear* functions: such definition gives us the capability to represent a wide variety of possibility distributions, with an eye to the implementation of the operations and to future extensions. A continuous membership function over the real domain is thus composed by *segments*, as follows:

```
class continuous_real_membership_function
      inherit Object public
      type list(segment);
```

where

```
type segment: tuple(x1: real,
                    y1: real,
                    x2: real,
                    y2: real,
                     extremes: integer);
```

O2FOOD Logical Structure

Elements for the definition of vague attributes

type segment;
classes discrete_integer_membership_function,
discrete_real_membership_function,
discrete_string_membership_function,
continuous_real_membership_function;

Functions that translate graphical operations into the appropriate sets of O2 command

NewClass, RemoveClass, NewAttribute, RemoveAttribute, NewCrispType, RemoveCrispType, NewVagueType, RemoveVagueType

Service functions for specific purposes

make_(norm_)trapezoidal, make_(norm_)triangular, make_(norm_)rectangular, make_(norm_)constant, segment_union, interpolate, intersection, view_function_graph.

Elements used to record graphical information

Figure 4: The logical structure of the O2FOOD schema.

A trapezoidal function is thus composed by at most three segments, a triangular function by at most two, a rectangular function and an isolated point by one segment; vertical segments, as well as the segments which lie on the horizontal axis, are not included in the segment list. The `extremes` parameter is used to discriminate between the inclusion or not of the extremes of the segment, and their possible extension to infinite.

By using the segments representation for membership functions, union, intersection and complementation operations always give as a result a membership function which is again representable by segments; observe that this closure property is not true if we consider only – for example – trapezoidal functions represented by means of the four parameters `a`, `b`, `c` and `d` as in Figure 5.

A facility that can be used to specify constant, rectangular, triangular and trapezoidal membership functions without the need to insert every segment is provided by the user interface, as explained in the next section. In particular, a (normalized) trapezoidal function can be specified by means of the four parameters `a`, `b`, `c`, `d`, as illustrated in Figure 5; the functions `make_xxx` and `make_norm_xxx` (where `xxx` stands for one of the words `constant`, `rectangular`, `triangular` or `trapezoidal`), contained into the O2FOOD schema, provide the correct segment representation.

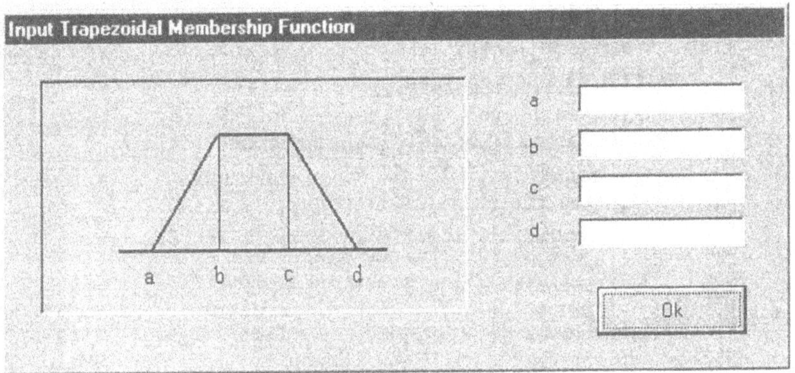

Figure 5: Trapezoidal membership functions can be inserted by specifying only the defining parameters `a`, `b`, `c` and `d`.

Even if the closure property of the set of trapezoidal functions with respect to the union, intersection and complementation operations is not satisfied in general, it may happen that the result of one of such operations is indeed a trapezoidal (or triangular, or rectangular, or constant) membership function. To recognize such cases, the methods `is_xxx()` have been designed; if the boolean value returned by one of these methods is `true`, the user can get the defining parameters `a`, `b`, `c`, `d` and `degree` by means of the functions `get_xxx(f)`, where `f` is a pointer to the object (instance of the `continuous_real_membership_function` class) containing the segments representing the membership function; the value returned is an appropriate `tuple` containing the cited parameters.

We conclude the discussion of the O2FOOD schema by observing that, as it can be seen in Figure 4, it contains also the following components:

- the functions that – exploiting the interaction with the `o2meta` schema – allow to add and remove classes, attributes and types in the conceptual schema maintained by O_2. Such functions can be thought as translators from the graphical operations performed by the user onto the graph to the appropriate sequences of O_2 commands. We will mention these functions when necessary, during the exposition of the features of the user interface;
- the data types that make possible to record the graphical information (position, size, associated labels, and others) of the elements that constitute the graph representing the conceptual schema onto the canvas.

6 The user interface

We turn now our attention to the user interface: we will briefly expose its operating features and our implementation choices.

When the user wants to develop a new conceptual schema the very first thing he/she has to do is to create a new *named system*. From a logical point of view, a named system is seen by O_2 as a collection of schemas with eventually some bases associated with them; when created, a named system is empty, i.e. it does not contain any schema or base. In order to use the features provided by the implementation of the FOOD model, the user has to prepare the newly created named system by *loading* the schemas O2FOOD and `o2meta` into it. Once these operations have been performed, the user interface has everything it needs to start: to access the elements of the conceptual schema under development the O2FOOD schema incorporates the pertinent meta-classes from the `o2meta` schema and automatically instantiates a meta-variable that provides the access to the methods of such classes.

At this point, the user can launch the `o2server` application by specifying the *named system*, and finally the client application by specifying the *named system* and a name for the schema that he/she wants to create or modify (*schema*):

`o2schema_designer -system` *named-system* `-schema` *schema*

If the indicated *schema* does not exist into the *named system*, a new empty schema and a base with the same name associated with it are created by means of the O2FOOD function `NewSchema`; on the other hand, if the named *schema* is found, its representation graph is loaded in memory and shown into the canvas, with each component exactly in the same position as it was left on the last session.

The base implicitly associated with the conceptual schema is used to hold both the instances of the classes contained into the schema, and the graphical information concerning the graph that represents the conceptual schema. To avoid possible inconsistencies between the definition of the conceptual schema (and of the graphic information of the respective graph) and the data stored into the base, the user is forced to use such implicit base to store the instances of the conceptual schema, and cannot specify a different base; moreover, modifications to the conceptual schema are not allowed when instances have already been inserted. An alternative solution, which is extendible to the multiple bases case (but is also computationally very expensive) would be to make the user interface invoke a function that checks for the consistency of every object with the definition of the corresponding class at every modification of the conceptual schema.

The information characterizing the graphical representation of the conceptual schema are stored into persistent collections of objects, one for each kind of graphical element; each collection is simply a `list` of `tuples`, where the attributes in each tuple contain the information about a single graphical element onto the canvas. We will not expose here the precise data structures we have used, as they have no theoretical interest; we have just mentioned them because they play a crucial role into the management of the visual part of the implementation, and they are subject to some non trivial processing during the session with the user interface.

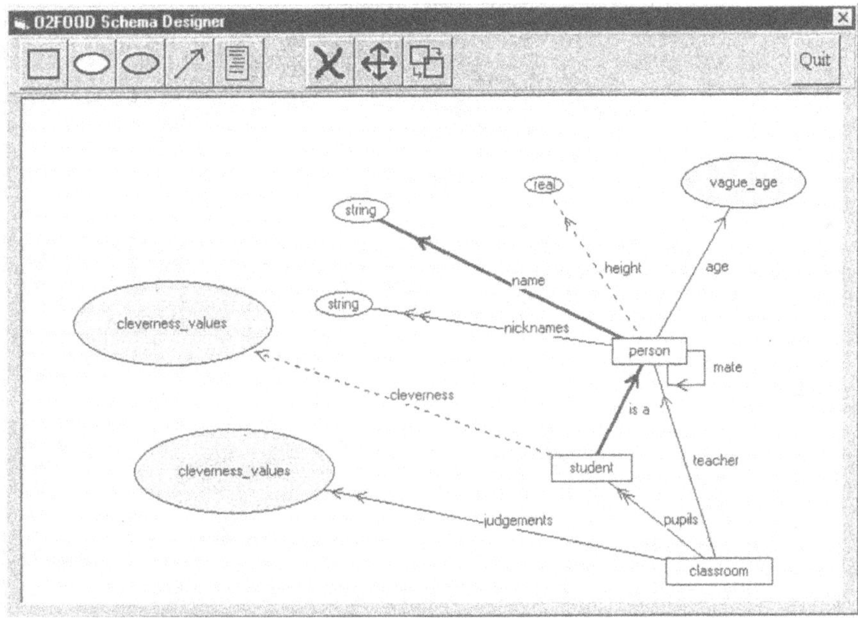

Figure 6: The main window, with an example of conceptual schema.

Let us now describe how the user interface is used to create and populate a conceptual schema. When it is launched for execution, the user interface presents a canvas with some *command buttons*, as illustrated in Figure 6. A command button for each category of graphical elements managed by the program is present, plus a button which allows to add/remove/edit the methods associated with a given class, a button which allows to delete a specified graphical element (and the associated schema element, through the invocation of appropriate commands to the O_2 system), a button for moving graphical elements around onto the canvas, a *clone* button which permits to have multiple copies of the desired graphical elements associated to crisp or vague types (to avoid muddle as much as possible) and, finally, the *quit* button which allows the user to leave the environment, after having made some checks about the existence of classes without attributes and/or methods, or the existence of (clones of) crisp/vague types without incoming arrows (that is, those elements that are not used as the type of any attribute in the conceptual schema).

The actions needed to design a conceptual schema are very intuitive: everything is done onto the graph, without the need to know the Data Definition Language; all the technicalities of schema construction are hidden by the program, thanks to the functions in the O2FOOD schema. In the following subsections, we briefly describe how the various elements of the conceptual schema are created, modified and deleted.

6.1 Creation of classes

Let us suppose that we want – for example – to define a new class `person` into the conceptual schema. What we have to do is nothing more than clicking on the class command button (upon which is drawn a rectangle, according to the FOOD model) and entering the name of the class (`person`, in this case). After having checked that a class or a type named `person` does not already exist, or that the user has specified a reserved word as the name of the class, the interface draws a rectangle containing the name of the class onto the canvas, and calls the `NewClass` function in the O2FOOD schema. Such function, through the methods provided by the `o2meta` schema, creates the empty class named `person` as follows:

```
class person inherit Object public
end class
```

Observe that we can consider the class `person` as being empty – that is, without any attribute or method – only from the FOOD model point of view; actually, the class is not empty at all because it inherits the generic methods (such as `edit()`, for example) defined for the class Object, the root of every class hierarchy in O_2.

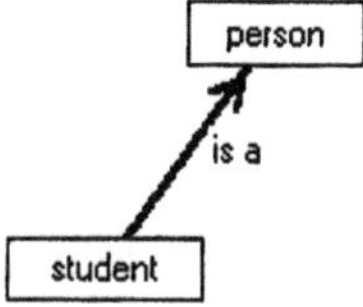

Figure 7: Graphical representation of the result of the creation of class `person` and its subclass `student`.

Now, if we want to define the class `student` as a subclass of `person`, what we have to do is simply to *select* the class `person` and perform the same actions we did to create that class. The graphical result is illustrated in Figure 7; in this case, the function `NewClass` of the O2FOOD schema defines the class `student` as:

```
class student inherit person public
end class
```

Each time the function `NewClass` creates a class into the conceptual schema, a new named constant is created too, to be used as a persistent collection to store the objects which are instances of the new class; thus we have, for example, the following named constant associated to the class `student`:

```
name student_objs: list(student);
```

6.2 Creation of crisp types

The addition of a *basic* crisp type (`integer`, `real`, `char`, `string` or `boolean`) to the conceptual schema is analogous to the creation of a class: it suffices to click on the crisp type command button (upon which is drawn an ellipse, according to the FOOD model) with the mouse and enter the crisp type name. An ellipse labeled with the name of the crisp type immediately appears on the canvas, and no command of the Data Definition Language is invoked to the O_2 system (basic types are predefined, and cannot be created or deleted).

In the case of a *structured* type, instead, after having clicked the crisp type command button and having entered the crisp type name, a check on the specified name is made and then the `NewCrispType` function of the O2FOOD schema is called; this function uses the standard O_2 *type editor* window (the same used by `o2tools`) which allows to specify the structure of the crisp type. Observe that, in this case, the knowledge of the Data Definition Language is required. When the structure of the crisp type has been specified, a `compile type` command is automatically invoked to the O_2 system, and a possible compilation error is reported to the user. When the compilation is finally successful, an ellipse appears onto the canvas, labeled with the newly created crisp type name.

6.3 Creation of vague types

The definition of vague types is one of the strongest points in the implementation of the FOOD model. To define a vague type, the user has to click on the vague type command button, upon which is drawn an ellipse with a gray background. After inserting the type name into a dialog window, the system performs the usual check on the name and then asks the user to specify the basic domain type for the membership functions that will be associated with the linguistic terms which constitute the possible values of a vague attribute. The possible choices for the domain type are shown in Figure 8. The next step is to define the set of values (linguistic terms) which can be assigned to a given attribute of that type. A new window – shown in Figure 9 – allows the user to insert, remove and modify such terms by means of the buttons `Add`, `Remove` and `Edit`; these operations vary according to the type of the membership function: as we have seen, in fact, the user will work with pairs (x, y) or with segments depending on the fact that the membership function is discrete or continuous and, in the first case the value of x is of type `integer`, `real` or `string`. Thus, if a discrete membership function has been specified, by pressing the `Add` button a window that allows the user to insert and/or remove the (x, y) pairs that compose the membership function will appear. In the case of continuous membership functions over the real domain, the window in Figure 12 prompts the user to specify the kind of membership function that he/she wants to associate with the linguistic term. Constant, rectangular, triangular or trapezoidal functions can be specified with the usual parameters `a`, `b`, `c` and `d`: a call to the appropriate function `make_norm_xxx` contained in the O2FOOD schema will provide the correct segment-based representation. By choosing the `Segments` option in Figure 10, the user can directly specify a general piecewise-linear function by using the input form shown in Figure 11.

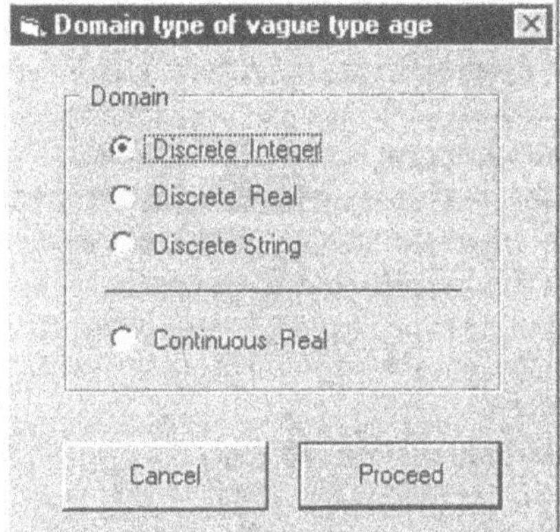

Figure 8: Possible choices for the basic domain of a vague type.

When a linguistic term is defined, it is added to the list of terms for the vague type under definition, as shown on the left of Figure 9. On the right of the window pictured in the same Figure, all the membership functions associated with the terms are drawn; clicking on a linguistic term, the corresponding function is highlighted.

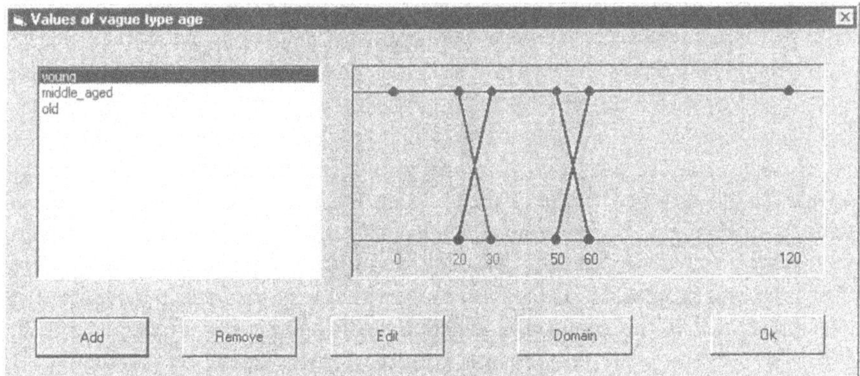

Figure 9: The window that allows the management of linguistic terms and the associated membership functions for vague types.

Another possibility given by the current implementation of the FOOD model is that the user can specify a *subrange* of a numerical domain when defining a vague type, as depicted in Figure 12. When the user chooses a subrange, he/she is not allowed to define pairs or segments that extend out of it. The button `Domain` shown in Figure 9 allows to change the previously chosen subrange for a defined vague type; if any of the membership functions associated with the linguistic terms extend over the new domain, the user is prompted to modify or delete them. Of course, it is not possible to change the domain type: for coherence, such change is not possible even in the case the user wants to convert a discrete integer domain into a discrete real domain.

When a vague type has been defined – by specifying the linguistic terms with their associated meaning (membership functions) – the user can always change its definition in the course of the session by selecting it and accessing to the window in Figure 9.

For the string domain, the definition of a vague type is slightly different; the difference is given by the peculiar nature of the domain, which has not a set of predefined values from which we can choose as happens for numerical domains. The user is thus prompted to insert the domain values into an appropriate window; when defining a membership function for a linguistic term, the user can specify the pairs (x, y) by choosing the value of x just from the list of available domain values (and the value of y in the interval [0,1]). In the portion of window that shows the membership functions associated with the linguistic terms of the current vague type (Figure9), domain values are shown in the order used to insert them. In this context, changing the string domain means adding or removing elements from the list of the actual domain values; when a string domain is changed, a check over the existing membership functions is done as happens for numerical domains.

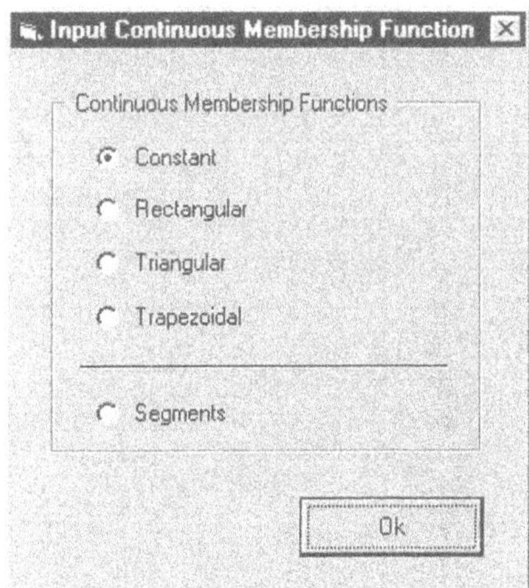

Figure 10: In the case of continuous real membership functions, the user can directly specify constant, rectangular, triangular and trapezoidal functions, or any piecewise-linear function by inserting a sequence of segments.

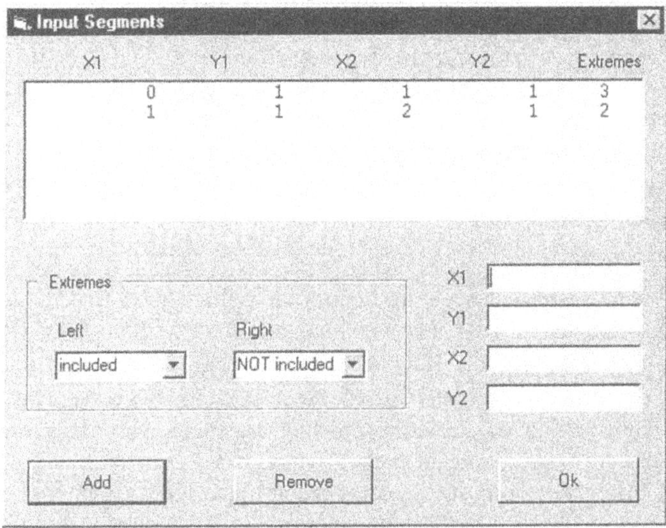

Figure 11: The input form used to specify the segments which compose piecewise-linear membership functions.

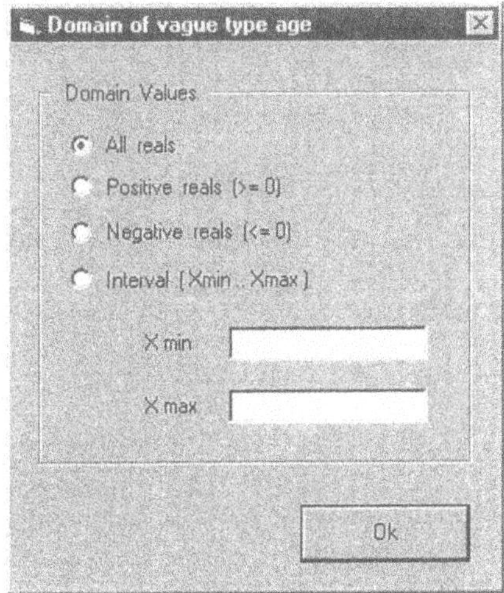

Figure 12: The user can choose to define a vague type over a *subrange* of the basic domain.

In all the cases, when we have specified the linguistic terms that constitute – together with the corresponding membership functions – the definition of a new vague type, the `NewVagueType` function from the O2FOOD schema is called; such function adds the definition of the new vague type to the conceptual schema, and creates a named object to store the actual membership functions in the implicit base associated with the schema.

6.4 Creation of attributes

As we have seen in section 5.1, when a class is created it does not contain any attribute; as a consequence, we need a procedure that allows to add attributes to a given class. Let us suppose that we want to add an attribute `height` of type `real` to the class `person`. Having created the class `person` and the crisp type `real` as described respectively in sections 5.1 and 5.2, we can select the class `person` by clicking on the corresponding rectangle with the mouse and by pressing the attribute command button, which has an arrow drawn upon it (see Figure 6). After indicating the name of the attribute we want to create, a pop-up window asks us the *kind* of attribute we desire, as illustrated in Figure 13; when we have made our choices, it suffices to click onto the canvas on the ellipse corresponding to the crisp type `real` to make appear an arrow that goes from the class to the specified type. The same procedure applies if the type of the attribute is not crisp: that is, if it is a vague type or a class (perhaps the class `person` itself). The arrow style depends upon the choices we have previously made, according to the FOOD model. In Figure 6 we can see some examples of crisp, vague and class attributes.

Before drawing the arrow onto the canvas, the O2FOOD function `NewAttribute` adds the newly defined attribute to the indicated class on the schema managed by the O_2 system. The data structure used to represent the new attribute varies depending on the characteristics specified in Figure 13; when needed, such structure contains the fields that allow to store – during the insertion of the instances of the class – the uncertainty degree (that, in a multiple attribute, applies to the entire m-uple) and the strength qualifier(s) in *T(Strength)* defined in (1) (that, in a multiple degree, apply to every element of the m-uple). Multiple attributes have been implemented through the `sets` construct of the O_2 Data Definition Language.

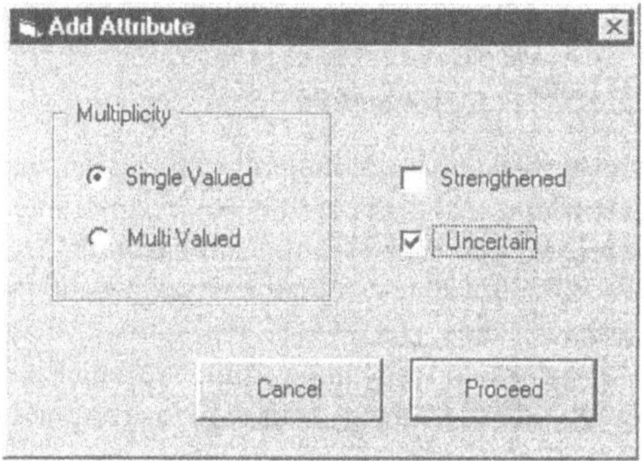

Figure. 13: The user can specify all the characteristics of an attribute described in the FOOD model.

Of course, the user interface provides also the capability to specify a precise value for a vague attribute; the data structure provides thus two additional fields, which store respectively the precise value and a boolean flag that indicates whether the meaning of the linguistic term is stored into the precise value field or into the membership function field. Observe that the possibility to insert a precise value into a vague attribute is just a facility, since precise values can always be represented through singleton membership functions. On the other hand, we point out that the conversion from precise values to the corresponding singleton membership functions is needed when evaluating a vague query; as a consequence, when the mechanism evaluating vague queries will be implemented, a new O2FOOD function that performs such conversion will have to be designed.

For aesthetic reasons, we have decided to limit to three the number of attributes that can go from a given class to another (or to itself); this limitation does not hold for crisp or vague types because we have the possibility to *clone* them, thus reducing the muddle produced by a large number of arrows.

6.5 Cloning crisp and vague types

To reduce the number of overlapping arrows into the graph representing the conceptual schema, the user has the possibility to clone crisp and vague types. Cloning a crisp or vague type is very easy: it suffices to select the desired type onto the graph and to press the *clone* command button in the main window (see Figure 6).

The cloning operation consists in creating an identical copy of the graphical element of the selected type; we point out that only the graphical element is duplicated, and not the definition of the type. As a consequence, there is absolutely no difference between the "original" graphic element and the clone: no relation occurs between the two elements, other than being a graphical representation of the same type.

The presence of clones complicates a bit the removal operation of a given type: in particular, the type definition is deleted only when the *last* corresponding graphical element is removed from the canvas (provided that the indicated type is not a crisp basic type); otherwise, only the graphical element is removed.

6.6 Moving graphical elements around

Cloning crisp and vague types is not the only way we can use to limit the number of overlapping arrows; to avoid this possibility further, every node of the graph can be *moved* by selecting it, by pressing the *move* button (which is depicted with four outgoing arrows, as illustrated in Figure 6) and finally by clicking on the new position on the canvas. Only classes and types (both crisp and vague) can be moved around; the arrows corresponding to the attributes are "anchored" to their starting and ending points, so they cannot be moved independently from the corresponding classes and/or types.

6.7 Management of methods

Methods, together with attributes, are the key components of classes; it is thus important to provide an efficient way to create, remove and modify them. Such operations are easily performed with the user interface, by means of the window shown in Figure 14, which is reached by selecting the desired class and by pressing the *method* command button in the main window (upon which it is drawn a written page, as illustrated in Figure 6).

As it can be seen from the Figure, the methods window contains the list of the methods currently defined for the specified class, as well as the buttons `Add`, `Remove` and `Edit`, by means of which the user can operate on such methods. The management of methods in the current implementation is completely based upon the tools provided by O_2 (i.e., the `o2tools` method editor). So, to write a method for the currently selected class, the user just presses the `Add` button and enters a name for the method; at this point, the `NewMethod` function from the O2FOOD schema invokes the O_2 editor – which allows to enter the head and the body of the method – and automatically invokes a `compile method` command when the user exits from the editor, reporting the possible error conditions. A completely analogous procedure is followed to modify (`Edit`) a method, while to remove an existing method it is sufficient to select its name and press the button `Remove`.

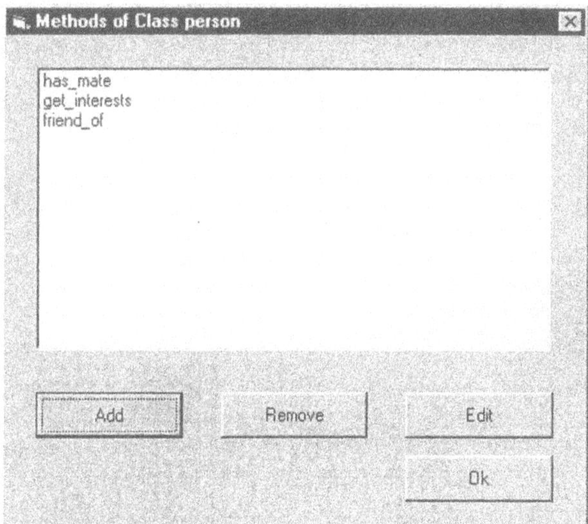

Figure 14: The window used to manage the methods of classes.

6.8 Removal of the elements of the schema

To remove a given graphical element, the user first selects it and then presses the *remove* button (which has an X drawn upon it, as shown in Figure 6).

While it is always possible to remove an attribute from a class, there are situations where a class itself or a type cannot be removed:

- when the class or type is used as the domain of values for an attribute; in particular, if the attribute goes from a class to the same class, the user must remove the attribute first;
- when the class is the superclass of another class in the schema.

If the user tries to remove a class that contains at least one attribute or method, or a structured crisp type, or also a vague type (that is, if he/she tries to delete something that required some effort to be defined), a warning message is shown. On the other hand, if the element to be removed is a clone of a crisp or vague type then no warning message is issued; moreover, the function `DeleteCrispType` (or `DeleteVagueType`) from the O2FOOD schema is called only when the *last* clone of a type is removed from the canvas, and the type itself is not a *basic* crisp type. As a last observation, when a class or vague type is deleted the corresponding named variable – that contains, respectively, the set of instances and the membership functions associated with the linguistic terms – is deleted too.

6.9 Management of the schema's instances

The last feature of the user interface is to allow the user to load the instances of the classes defined into the conceptual schema. To do this, the user has just to click twice on the rectangle corresponding to the interested class; a window containing the list of inserted instances appears. Every instance is identified by a "nickname" (indeed, an identifier), and the creation, removal and modification of the instances are performed through the buttons `Add`, `Remove` and `View/Modify`, as happens in the management of methods.

Appropriate entry forms are shown when the user requests to load or view the value for a given attribute, depending on its type. Since the type of an attribute can be a class, the loading procedure – as well as the view procedure – is recursive. When an attribute of type class is encountered during the loading of values, the user has two possible choices: to create a new instance of the specified class or to load the value `null`; loading the value `null` is, in particular, the only way to end the loading procedure when we encounter a class with an attribute pointing to itself. When a vague type is encountered, the user has also in this case two possibilities: to choose one of the linguistic terms defined for the vague type or to specify a precise value from the underlying domain – that is, a numerical value or a string – as explained in section 5.3. For structured crisp types, the values are loaded through the invocation of the `o2look edit()` method on the whole attribute. For uncertain attributes the uncertainty degree is requested, while for strengthened attributes the user has to choose among one of the predefined strength qualifiers presented into a list.

Finally, if the attribute is multiple, a sequence of values is requested, terminated by pressing the `End` button into the dialog window; in such a case, if the attribute is also strengthened then for every element of the sequence a strength qualifier is requested to the user, while if the attribute is uncertain a single uncertainty degree for the entire *m*-uple is requested when the user presses the `End` button.

6.10 Ending the session with the user interface

The last command button in the main window of the user interface is the *quit* button (see Figure 6). We observe that it is unnecessary for the user to *save* the modifications made to the conceptual schema, because every change has an immediate effect into the schema maintained by the O_2 system. As a drawback, no undo operation is possible; a rough escape to this situation is to make a copy of the schema and the associated base before launching the client, and reverting to this copy if the modifications are not satisfying.

The quit operation, however, does not simply end the session by disconnecting the client from the O_2 server. Because there is always the possibility that an absent-minded user defines a class or type without ever using it, a check is made by the program to avoid this possibility. The graph is then visited looking for classes without attributes and methods or types without incoming arrows: when one of these elements is found, the user is asked whether such element has to be kept or not. Different marks – recorded together with the graphical information – are used to distinguish either the elements in the graph that can/cannot be removed, or the unused elements that the user has nonetheless decided to keep. Observe that multiple visits to the graph may be necessary: to convince ourselves of this fact, it suffices to think to a chain of `is-a` relations made up by empty classes. Inner classes cannot be removed without breaking the chain, so the user is prompted about the last class only; if the user decides to remove such class, then he/she would have to be prompted about the next class in the chain, but there is the possibility that such class has already been visited and marked as non-removable.

7 Future developments and open issues

The first thing we must point out is that the current implementation of the FOOD model is still a prototype, for it lacks many of the characteristics which are usually found in a full-featured application; as an example, the management of graphic elements with the mouse is performed in an easily-codable fashion, so that it lacks the possibility to select more than one object at a time, to drag the objects, to collapse or explode parts of the graph, and so on.

One limitation of the current implementation is that the user cannot specify as a value of a class attribute an existing instance. A possible solution would be to duplicate the indicated instance (identified by means of its nickname) to allow the user to specify different values for the uncertainty degree and for the strength qualifier(s); from that moment on, however, the two instances would be modifiable independently one from the other.

Another limitation is that the user cannot modify the conceptual schema when he/she has already populated it with some objects. As we have told in section 5, to avoid inconsistencies between the schema definition and its instances, after each schema modification it would be necessary to scan the entire set of objects; because a given object can be part of another object, if we modify a class then the scan of the set of instances of that class is not enough.

A further weak point of the implementation is in the management of methods: when a crisp or vague type is deleted, no check over the use of such type by the methods present in the conceptual schema is performed, so there is the possibility to have some methods which raise an execution error when invoked due to the lack of one of the required types.

The whole graphical implementation of the user interface can be improved: in particular, in the current implementation we have decided not to represent the instances of the schema as graphs, as the FOOD model would require.

Finally, the current implementation of the FOOD model does not allow to define fuzzy classes and fuzzy hierarchies.

On the other hand, we have particularly taken care of the software design aspects. As we have already observed the implementation has been conceived in a completely open fashion; to better support future extensions, some functionalities contained in the O2FOOD schema have been implemented even if they are not used in the current implementation, such as the possibility to shift membership functions defined on a numerical domain (useful, for example, to define the semantics of strength qualifiers such as *very* or *scarcely*), or the possibility to perform the the fuzzy pattern matching operations. Moreover, as we have already pointed out, if changes and enhancements concern only the graphical interface, these can be first tested with the Visual Basic prototype.

References

1. G.Bordogna, D. Lucarella, G.Pasi, "An extension of a graph based data model to manage fuzzy information", *in "Fuzzy and Uncertain Object-Oriented Databases"*, R. De Caluwe ed., World Scientific, London, 97-122, (1997).
2. G.Bordogna, G.Pasi, D. Lucarella "A Fuzzy Object Oriented Data Model Managing Vague and Uncertain Information", *International Journal of Intelligent Systems,* 14(6), (1999).
3. Motro, "Imprecision and Uncertainty in Database Systems", in: *Fuzziness in Database Management Systems, P. Bosc, J. Kacprzyk (eds.),* Physica-Verlag, Heidelberg, 3-22 (1995).
4. L.A.Zadeh, "Fuzzy Sets as a Basis for a Theory of Possibility", *Fuzzy Sets and Systems*, 1, 3-28 (1978).
5. D. Dubois, H. Prade, *Possibility Theory: an Approach to Computerized Processing of Uncertainty*, Plenum Press, New York, (1988).
6. M.Gyssens, J.Paradaens, D.Van der Gucht, "A Graph-Oriented Object Database Model", *IEEE Trans. on Knowledge and Data Engineering,* 6(4), (1994).
7. D.Lucarella, A.Zanzi, "A Graph-Oriented Data Model", in: *Database and Expert Systems Applications* R.Wagner, H.Toma (Eds.), Springer Verlag, Berlin, 197-206 (1996).
8. G.Bordogna, D. Lucarella, G. Pasi, "Graph Based Interaction in a Fuzzy Object Oriented Database", in *"New Trends in Fuzzy Logic II, proc. of* WILF'97", *P.Blonda, M.Castellano, A.Petrosino eds.,* World Scientific, 115-128, (1997).
9. P.Bosc, O.Pivert, A new approach to the filtering of ill-known data, *Proc. of the 7th IEEE Int. Conf on Fuzzy Systems (FUZZ-IEEE'98)*, Anchorage (USA), 1302-1307, (1998).
10. O.Deux et al., "The O2 System", *Communications of ACM*, 34(10), 34-48 (1991).
11. D. Dubois, and H. Prade, Tolerant Fuzzy Pattern Matching: an Introduction, in Fuzziness. In: *Database Management Systems*, P. Bosc and J. Kacprzyk eds., Physica Verlag, 42-59, (1995).
12. P.Bosc, H. Prade, "An Introduction to the Fuzzy Set and Possibility Theory-based Treatment of Flexible Queries and Uncertain or Imprecise Databases", in: *Uncertainty Management in Information Systems, A. Motro, P. Smets (eds.),* Kluwer Academic Publishers, 285-324 (1997).
13. R.G.G.Cattell. *The Object Data base standard: ODMG-93*. Morgan Kaufman, San Mateo, CA (1993).

Studies in Fuzziness and Soft Computing

Vol. 25. J. Buckley and Th. Feuring
Fuzzy and Neural: Interactions and Applications, 1999
ISBN 3-7908-1170-X

Vol. 26. A. Yazici and R. George
Fuzzy Database Modeling, 1999
ISBN 3-7908-1171-8

Vol. 27. M. Zaus
Crisp and Soft Computing with Hypercubical Calculus, 1999
ISBN 3-7908-1172-6

Vol. 28. R.A. Ribeiro, H.-J. Zimmermann, R.R. Yager and J. Kacprzyk (Eds.)
Soft Computing in Financial Engineering, 1999
ISBN 3-7908-1173-4

Vol. 29. H. Tanaka and P. Guo
Possibilistic Data Analysis for Operations Research, 1999
ISBN 3-7908-1183-1

Vol. 30. N. Kasabov and R. Kozma (Eds.)
Neuro-Fuzzy Techniques for Intelligent Information Systems, 1999
ISBN 3-7908-1187-4

Vol. 31. B. Kostek
Soft Computing in Acoustics, 1999
ISBN 3-7908-1190-4

Vol. 32. K. Hirota and T. Fukuda
Soft Computing in Mechatronics, 1999
ISBN 3-7908-1212-9

Vol. 33. L.A. Zadeh and J. Kacprzyk (Eds.)
Computing with Words in Information/Intelligent Systems 1, 1999
ISBN 3-7908-1217-X

Vol. 34. L.A. Zadeh and J. Kacprzyk (Eds.)
Computing with Words in Information/Intelligent Systems 2, 1999
ISBN 3-7908-1218-8

Vol. 35. K.T. Atanassov
Intuitionistic Fuzzy Sets, 1999
ISBN 3-7908-1228-5

Vol. 36. L.C. Jain (Ed.)
Innovative Teaching and Learning, 2000
ISBN 3-7908-1246-3

Vol. 37. R. Słowiński and M. Hapke (Eds.)
Scheduling Under Fuzziness, 2000
ISBN 3-7908-1249-8

Vol. 38. D. Ruan (Ed.)
Fuzzy Systems and Soft Computing in Nuclear Engineering, 2000
ISBN 3-7908-1251-X

Vol. 39. O. Pons, M.A. Vila and J. Kacprzyk (Eds.)
Knowledge Management in Fuzzy Databases, 2000
ISBN 3-7908-1255-2

Vol. 40. M. Grabisch, T. Murofushi and M. Sugeno (Eds.)
Fuzzy Measures and Integrals, 2000
ISBN 3-7908-1255-2

Vol. 41. P. Szczepaniak, P. Lisboa and J. Kacprzyk (Eds.)
Fuzzy Systems in Medicine, 2000
ISBN 3-7908-1263-4

Vol. 42. S. Pal, G. Ashish and M. Kundu (Eds.)
Soft Computing for Image Processing, 2000
ISBN 3-7908-1217-X

Vol. 43. L.C. Jain, B. Lazzerini and U. Halici (Eds.)
Innovations in ART Neural Networks, 2000
ISBN 3-7908-1270-6

Vol. 44. J. Aracil and F. Gordillo (Eds.)
Stability Issues in Fuzzy Control, 2000
ISBN 3-7908-1277-3

Vol. 45. N. Kasabov (Ed.)
Future Directions for Intelligent Information Systems on Information Sciences, 2000
ISBN 3-7908-1276-5

Vol. 46. J.N. Mordeson and P.S. Nair
Fuzzy Graphs and Fuzzy Hypergraphs, 2000
ISBN 3-7908-1286-2

Vol. 47. Ernest Czogała† and Jacek Łęski
Fuzzy and Neuro-Fuzzy Intelligent Systems, 2000
ISBN 3-7908-1289-7

Vol. 48. Masatoshi Sakawa
Large Scale Interactive Fuzzy Multiobjective Programming, 2000
ISBN 3-7908-1293-5

Vol. 49. Ludmila I. Kuncheva
Fuzzy Classifier Design, 2000
ISBN 3-7908-1298-6

Vol. 50. Fabio Crestani and Gabriella Pasi
Soft Computing in Information Retrieval, 2000
ISBN 3-7908-1299-4

Vol. 51. János Fodor, Bernard De Baets and Patrice Perny (Eds.)
Preferences and Decisions under Incomplete Knowledge, 2000
ISBN 3-7908-1303-6

Vol. 52. Etienne E. Kerre and Mike Nachtegael (Eds.)
Fuzzy Techniques in Image Processing, 2000
ISBN 3-7908-1304-4

www.ingramcontent.com/pod-product-compliance
Ingram Content Group UK Ltd.
Pitfield, Milton Keynes, MK11 3LW, UK
UKHW021829190726
13853UKWH00003B/1259

* 9 7 8 3 6 6 2 0 0 3 7 9 4 *